Conceptual Design of Multichip Modules and Systems

Conceptual Design of Multichip Modules and Systems

Peter A. Sandborn
Hector Moreno

Microelectronics and Computer Technology Corporation (MCC)

KLUWER ACADEMIC PUBLISHERS
BOSTON/DORDRECHT/LONDON

Distributors for North America:
Kluwer Academic Publishers
101 Philip Drive
Assinippi Park
Norwell, Massachusetts 02061 USA

Distributors for all other countries:
Kluwer Academic Publishers Group
Distribution Centre
Post Office Box 322
3300 AH Dordrecht, THE NETHERLANDS

Library of Congress Cataloging-in-Publication Data

Sandborn, Peter A., 1959 -
Conceptual design of multichip modules and systems / Peter A. Sandborn, Hector Moreno.
p. cm. -- (The Kluwer international series in engineering and computer science ; SECS 250)
Includes bibliographical references and index.
ISBN 0-7923-9395-3 (alk. paper)
1. Multichip modules (Microelectronics)--Design and construction. I. Moreno, Hector, 1942 - . II. Title. III. Series.
TK7870.15.S26 1994
621.381'046--dc20 93-32222
CIP

Printed on acid-free paper.

Printed in the United States of America

Contents

Preface

This book treats activities which take place at the earliest stages in the design of multichip systems. These activities include the formalization of design knowledge (information modeling), tradeoff analysis, partitioning, and the process of capturing design decisions (information dynamics), all of which take place prior to the traditional CAD activities of physical design. The need for this book arises from increasing pressure on designers to ensure first pass success in the creation of complex multichip systems. Conceptual design is at the forefront of this need, because 80% of a product's final cost is determined by decisions made in the first 10% of the design cycle. While the philosophies and approaches addressed in this book have been touched upon in recent books focusing on the design of integrated circuits, to the authors' knowledge, no texts which treat conceptual design from a packaging and interconnection viewpoint exist today.

One of the most difficult and frustrating problems in multichip system design is the concurrent management of a large number of interdisciplinary performance constraints and requirements. Most engineers have become specialized in a single view of the problem and are not well equipped to balance highly technical design concerns against economic and manufacturing realities. Packaging and interconnect design requires designers who are experts in everything from electrical modeling, to design-for-test, to cost modeling.

This book makes a special effort not to focus exclusively on Multichip Modules (MCMs). While a multichip system may be a multichip module, more traditional packaging approaches such as single chip packages, which are through-hole or surface-mounted to printed circuit boards, also represent multichip systems. While MCMs may provide the best high performance, high density packaging solution for some applications they do not represent the optimum solution for many other applications at this time, nor is it likely that they will provide the best tradeoff for every problem in the future. Furthermore, optimal solutions usually consist of mixtures of technologies and can rarely be completely determined from simple generalizations. We therefore have adopted the term "multichip system" to refer to MCMs and traditional single chip package approaches.

We intend for this book to appeal to a broad range of interests, including packaging and interconnect design, concurrent engineering, computer aided design, and system synthesis. The book is useful both as a reference for system designers as well as a text for students wishing to gain a perspective on the nature of packaging and interconnect design. We did not write this text as a review or reference on packaging and interconnect or multichip module technologies. In our opinion several adequate works exist now and there will be no lack of new texts appearing in the future.

The chapters is this book form a coherent unit, but they are structured so that any one chapter is sufficiently self-contained to be read independently from the rest. Chapter 1 describes and defines the area of conceptual design and discusses the role that conceptual design plays in a multichip systems design environment. Chapter 2 focuses on the information which is required for conceptual design and how that information is modeled. Chapter 2 also addresses the process of capturing the design decisions. Chapter 3 discusses how tradeoff analysis is performed and contains detailed discussions and derivations of metrics used to estimate the performance of the systems under design. Chapter 4 introduces physical and functional partitioning of multichip systems. Chapter 5 provides example tradeoff analyses of multichip systems performed using the techniques developed in Chapter 3 and 4.

The authors are fortunate to work for the Microelectronics and Computer Technology Corporation (MCC) through which they can tap the knowledge of a large number of participating companies. We would especially like to thank those companies who sponsored the catalyst for the development of the knowledge contained in this book. It is the members of the Multichip Systems Design Advisor project at MCC who had the foresight to realize the significance of the conceptual design problem. In addition we wish to thank the Advanced Research Projects Agency (ARPA) for their support of MCC's information modeling work.

We wish to thank our colleagues at MCC: Magdy Abadir, Linda Bal, Ken Drake, Rajarshi Ghosh, Ashish Parikh, Tom Dolbear, Hassan Hashemi, Claude Rathmell, Bill Weigler, and David Clegg who have contributed their expertise to this subject, making our job easier, and to Larry Smith and Jon Prokop for providing detailed reviews of the manuscript. Finally, we would like to acknowledge the enthusiastic support of our publisher, Ken Tennity at Kluwer; and a special thanks to our wives and children for their love, support, and patience throughout this project.

P.A.S.
H.M.
Austin, Texas

Conceptual Design of Multichip Modules and Systems

Chapter 1

Introduction

1.1 Overview of Microelectronic Packaging

Microelectronic packaging is the science of providing interconnections and a suitable operating environment for microelectronic circuits [1.1]. Packaging involves the critical activities of interconnecting, powering, cooling, and protecting semiconductor chips. The technologies, and materials available for microelectronic packaging are numerous. Figure 1.1 diagrams the electronic module assembly schemes in use today. All methods begin with chips (also referred to as bare die) and conclude with a completed module which could be interconnected within a larger system. The multiple levels shown in Figure 1.1 are often referred to as the packaging hierarchy. The number of levels within the hierarchy varies depending on the module assembly path chosen. Figure 1.1 does not show the complete packaging hierarchy. Traditionally, the complete hierarchy could include other elements such as a board or backplane level to which modules or cards are connected, a cable level (sometimes called gate level) in which boards are connected by cable, and a frame or box level which contains the entire system [1.1].

The following subsections briefly review the most common microelectronic packaging and assembly approaches. It is not the intent of this book to provide a detailed technology review. Detailed technology overviews are provided in several other texts [1.1,1.2,1.3].

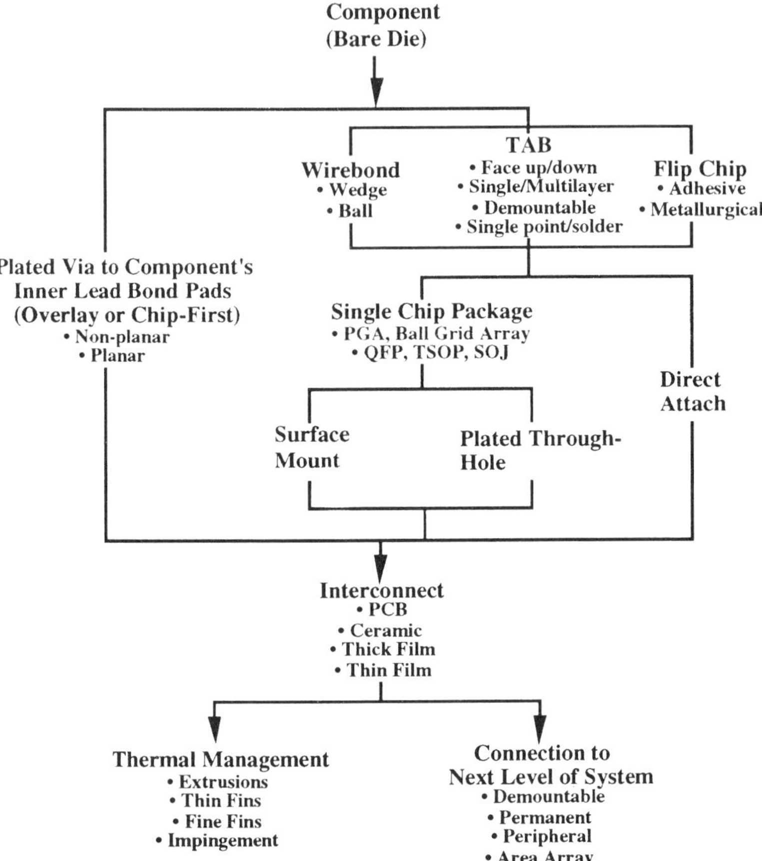

Figure 1.1. Module assembly schemes. Chips can be attached using wirebond, TAB (Tape Automated Bonding), or a flip chip method into a single chip package or directly on the module (interconnect). An alternative approach in which bare die are placed into cavities machined into a substrate and the interconnect fabricated on top (overlay or chip-first) is also possible. Single chip packages are usually referred to as the first level in the packaging hierarchy and the interconnect is referred to as the second level. Methods which attach the chip directly to the interconnect skip the first level packaging. Also included within this module-level packaging hierarchy are cooling approaches and connectorization to the rest of the system.

1.1.1 Single and Few-Chip Packages on Printed Circuit Boards

This assembly approach involves the placement of the die into any of several types of single chip packages including Dual In-line Packages (DIPs), Quad Flat Packages (QFPs), Pin Grid Arrays (PGAs), Thin Small Outline Packages (TSOPs), Small Outline J-lead packages (SOJ), and Ball Grid Arrays (BGAs). The die are generally attached to a leadframe using an adhesive, and the die I/O are electrically connected to bond pads on a leadframe using wirebonding. TAB and flip chip bonding of die into single chip packages is also possible but less common at this time. The bonded die is then molded for protection, and pins or solder bumps are added, or leads which are part of the lead frame are bent into position. The single chip package is next connected to a printed circuit board using either through-hole mounting or surface mounting. Figure 1.2 shows a printed circuit board with through-hole

Figure 1.2. Example of Printed Circuit Board (PCB) packaging. All components on the board shown are surface mounted.

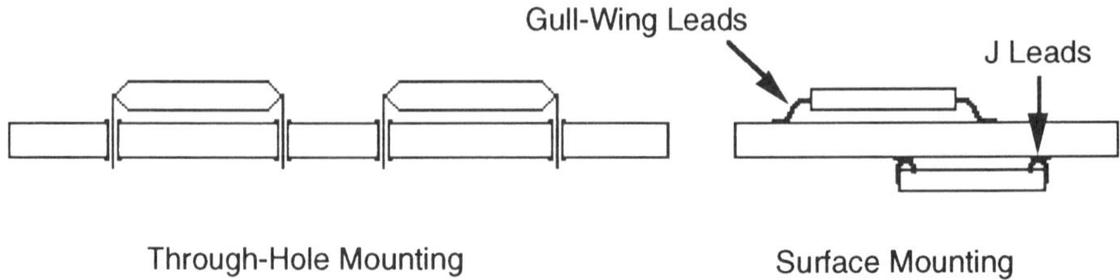

Figure 1.3. Through-hole and surface mounting on a printed circuit board (PCB).

and surface mounted components on it.

Though-hole mounting of single chip packages has been the traditional low-cost packaging choice and remains the most widely used packaging type today (Figure 1.3). Pins which are attached to the bottom or sides of the single chip package are mounted in holes drilled in the printed circuit board. Printed circuit boards, designed for through-hole mounting, use the plated through-holes drilled in the board to make connections between wiring levels. Once the pins of the single chip packages or sockets are inserted through the holes, the bottom surface of the board is dipped in a solder solution that adheres only to the pins, plated walls of the holes, and the contact areas surrounding the holes. Reflow soldering refers to methods in which solder is applied to parts before they are joined, then the parts are heated to cause the solder to remelt or reflow.

While through-hole technology is mechanically sound and very reliable, the through-holes often make mounting chips on both sides of a board impractical and increase the space needed for lead clearance on the back of the board. Through-hole mounting also becomes expensive as the number of I/O per device becomes large because manual (instead of machine) placement of the packaged chips on the board becomes necessary. A detailed treatment of through-hole mounting can be found in [1.4].

A more advanced printed circuit board attach technology called surface mounting solves many of the shortcomings of through-hole technology (Figure 1.3). In the surface mount approach a single chip package is soldered directly to the surface of a board without requiring any through-holes for mounting (via holes are still present in surface mount boards for interconnection between layers). Single chip packages for surface mount applications can be leadless with metallization on the sides of the packages, leaded, or bumped for reflow soldering. The smaller outline packages, lack of through-holes, and possibility of mounting chips on both sides of the board generally improve packaging density over through-hole techniques. In addition the surface mount manufacturing process is easier to automate because placing packages on the surface of a printed circuit board is faster and simpler than forcing pins through holes. However, testing of surface mount boards for electrical and mechanical defects, is more difficult than for through-hole implementations and is complicated further when components are placed on both sides of the board. Thermal mismatch problems are also more severe when using surface mount than through-hole mounting. A detailed treatment of surface mount can be found in [1.5].

All techniques which involve the use of single chip packages inherently provide some measure of chip protection, easier handling, and less costly test and burn-in of the chips than bare die mounting approaches.

Traditionally, only one die is placed in a sealed package for surface or through-hole mounting. Few-chip packaging involves the placement of more than one die into a sealed package. Several approaches have been used for few-chip packages: multisite or multichip leadframes and small multichip modules (few-chip modules) placed inside of traditional ceramic or plastic packages. Multichip tape has been fabricated using a multilayer TAB tape process. This process uses a multichip leadframe onto which more than one chip can be bonded. Multichip leadframes have also been fabricated by modifying conventional single chip leadframes.

Numerous efforts have focused on placing small multichip modules into traditional single-chip packages. In some cases small printed circuit boards containing multiple chips have been attached to the die attach pad of a leadframe for packaging into a single chip package. Similarly, silicon-on-silicon modules containing multiple chips have been placed inside plastic chip carriers, and into ceramic PGA's. Figure 1.4 shows an example few-chip package.

Figure 1.4. An analog crossbar switch implemented on a silicon-on-silicon interconnect. The module is wirebonded into an 88 pin ceramic PGA package [1.6]. (Photograph courtesy of PMC Sierra)

The use of few-chip packages generally increases the packaging density over exclusively single-chip packaged systems and may lead to a decrease in the total assembly cost through automation and continuous processing, however, few-chip packages are also harder to test and repair.

1.1.2 Multichip Modules (MCMs)

Multichip modules completely eliminate one level of packaging, namely the single chip package. Direct die mounting places bare die directly on or in a board or module. Multichip modules can be constructed using a number of different thick or thin film processes and assembly schemes. Multichip modules are categorized as MCM-L: Laminated interconnect (fine line printed circuit board), MCM-C: Ceramic interconnect (Low Temperature Cofired Ceramic - LTCC, High Temperature Ceramic), and MCM-D: Deposited interconnect (thin and medium film).

A common MCM-L approach is Chip-On-Board (COB). The COB approach uses fine line multilayer printed circuit board technology with bare die directly attached to the board using TAB or wirebonding. Chip-On-Board is presently used primarily in consumer applications where chip I/O counts are small and weight and size are important. Employing only mild extensions of existing printed circuit board technology, COB can be very inexpensive. While COB can yield modules which are competitive with thin film techniques in cost, size, and speed, they are not always practical for applications which require through-substrate cooling because of their relatively thick dielectric layers. Through-substrate cooling can be accomplished at the expense of wiring capacity by including thermal vias or slugs below critical components.

Ceramic thick and thin film multichip module technologies (MCM-C) are highly reliable and well characterized. Ceramic interconnects are, however, less available and more expensive than printed circuit boards. Ceramic approaches generally utilize materials with coarse granularity (i.e. green sheets, glass-cloth) and involve processes which are not easily adapted to producing layers with micron dimensions. In low temperature cofired ceramic (LTCC) approaches, ceramic layers are punched, screened with ink and stacked together. Once all the layers are completed they are laminated together (i.e., cofired) and sawed to the appropriate size. Unlike printed circuit boards, where layers are usually laminated together and the holes drilled later, cofired ceramic interconnects have their via holes punched before the layers are fired together. This fabrication procedure results in via alignment which is basically independent of the number of layers and it is possible to obtain smaller vias than with common printed circuit board construction. Unlike the LTCC approach, thick film processes involve the sequential build-up of multilayers. Thick film approaches start with ceramic substrates and screen on metal and dielectric layers firing between each layer.

Two different assembly approaches have been used for thin film multichip modules (MCM-D): traditional placement of die on top of a completed interconnect (the "chip-last" approach) or placement of die in a machined substrate (not a leadframe) followed by fabrication of the interconnect over the top of the die (the overlay or "chip-first" approach). In the traditional assembly approach of placing die face-up on top of an interconnect, die are attached to a top bonding layer using an adhesive. The adhesives are usually epoxies or polyimides filled with thermally conductive particles. Gold-based eutectics and filled solder glass are also commonly used for die attach. After the die are attached to the interconnect, they are electrically connected to the bonding pads using wirebond or TAB. In the case of flip chip, the die attach and the electrical connection are one and the same. However, thermally conductive materials may be injected between the bonds or under the die. Figure 1.5 shows a thin film copper/polyimide multichip module built for the MCC Government Open Systems Project.

The alternative "chip-first" multichip module construction technique offers some advantages over the traditional approach. The basic

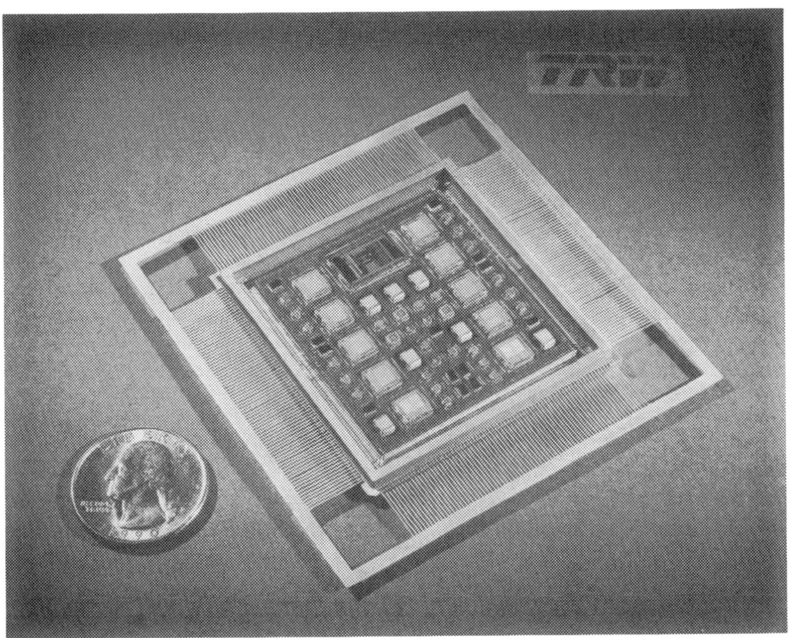

Figure 1.5. A TRW 50 MHz control memory module implemented in a thin film copper/polyimide multichip module built by Raytheon for the MCC Government Open Systems Project.

concept behind the overlay approach is the building of the interconnect on top of the chips. The equivalent of chip assembly occurs at the beginning of the interconnect fabrication when the die are glued into a specially machined substrate. A metal/dielectric interconnect is then fabricated sequentially over the top of the chips. In the General Electric version [1.7], the fabrication process optically locates and laser drills vias to the bond pads on the die essentially "skipping" the bonding step. Laser lithography is required to dynamically align and write the interconnect to the chip I/O. The overlay process offers the highest packaging density presently available at the expense of greater rework complexity if components require replacement after final module test.

A detailed review of thin film multichip module technology is contained in [1.8].

1.2 Conceptual Design

Conceptual design refers to module and system design activities that take place prior to detailed simulation activities and "physical" design (i.e., layout, placement, routing, verification, etc.). Conceptual design, as we are defining it, represents the activities necessary to create the specification for a system. The complete synthesis activity would include system architecture design, creation of hardware and software specifications, and the development of netlists and other information needed to begin physical design.

1.2.1 Understanding the Problem

One of the serious shortcomings in the present computer-aided design (CAD) tool sets for packaging and interconnect system level design is in the area of conceptual design support and specifically tradeoff analysis and partitioning. The need for higher system performance has brought about the requirement for advanced packaging and interconnection approaches which do not compromise the higher performance of advanced chips. Traditionally packaging and interconnect CAD tools have evolved from more mature tools used in IC design activities. This scenario is changing as designers realize that IC and printed circuit board tools can not evolve to support the unique problems posed in the multichip module design space. The design of complex modules and systems is characterized by a greater diversity of system and subsystem components (and assembly approaches) than chip-level design, requiring an interdisciplinary analysis, highly dependent upon physical and functional partitioning. Tradeoff analysis for these systems requires expertise in electrical, thermal, mechanical, manufacturing (including cost analysis), testing, and reliability analysis. This analysis is further complicated in system packaging and interconnect design by the presence of a large set of competing technologies, materials, and process alternatives.

For most system designers, the goal is to find a design which balances performance with ease of manufacture and support, while minimizing cost. Early in the design cycle, numerous tradeoff decisions have to be made intelligently and quickly before major investments are committed or detailed design work begins. Besides time and budget constraints, system packaging tradeoff decisions are also hampered by the lack of adequate and accessible alternative technology data. These critical decisions, made within the first 10% of the total design cycle time, will ultimately define up to 80% of the final product cost. Therefore, making appropriate choices early in the design cycle will have a significant impact throughout the design, production, and lifecycles.

The lack of a complete set of technology information is an especially serious problem in the packaging and interconnect field since the number of technologies, processes, materials, and approaches is substantial, and selecting optimums is arduous and non-trivial if one truly wants a balance in cost and performance. Alternative technologies and materials include: interconnects (printed circuit boards, ceramic, thin film, etc.), packaging materials, bonding techniques (wirebond, TAB, flip chip), test techniques, and manufacturing methods. The designer may not be aware of all the technology choices that exist, and few designers can comprehend all the interdependencies and ramifications the technologies and materials chosen may have on a particular design's cost and manufacturability. Furthermore, the refinement of the tradeoff analysis must continue throughout the design cycle, so that as more accurate information is obtained, there is a way to confirm that the selected technologies, processes, and materials are, in fact, still providing the "best" system solution.

The selection of an optimal combination of technologies depends on three drivers: 1) the characteristics of the components to be integrated, 2) the application for which the system will be used (i.e. performance requirements, cost, operating environment, and support requirements), and 3) the availability of previously designed structures and their design history. In addition, the problem is not confined to just the selection of packaging technologies for a module or just the selection of a chip technology. Complete tradeoff analysis must include chip/component through whole system design, and hence, must occur during all phases of the design process. The ideal tradeoff environment would allow the custom design of ICs for the packaging environment concurrent with packaging design and technology selection.

1.2.2 The Estimation-Simulation Gap

Tradeoff analysis activities could be carried out using detailed point solution simulation tools. However, because minimal information about the system is available during the conceptual design phase, the usability of detailed simulators is limited. In addition, the lack of a well-defined theory relating behavior and structure at the system-level often makes approaches based on detailed computations less favorable than knowledge-base supported estimation methods for conceptual system design and synthesis. Relying on detailed simulations to provide the information necessary to make system-level implementation decisions can be a risky, resource intensive undertaking. Providing intelligent assistance to the designer in the synthesis process often proves to be more useful than simulation in the early stages of design. Estimation-based conceptual tradeoff techniques are not intended to replace detailed simulation, but to provide support for early (conceptual) specification design when simulation is may not be practical.

Two approaches have helped close the gap between estimation and simulation: methodology managers for detailed simulation tools and automated model building for detailed simulation tools. The first approach consists of the development of tools which provide guidance and automatic data manipulation for the use of detailed simulation tools. The integration of various modeling tools under a single user interface does not, by itself, make detailed point solutions practical at the conceptual level. However, adding elements of methodology management can make their use more effective during conceptual design. An existing tool that operates as a simulation manager is the Packaging Design Support Environment (PDSE) [1.9]. The PDSE integrates stand-alone parameter extraction modeling tools for simulating transmission lines.

Besides the management or guidance of point solution tools, the required tool inputs need to be made compatible with the information available at the conceptual level of design. The second approach is to automatically build models for use in detailed simulators and post-process their results back into the conceptual design space. Several examples of software tools which construct models for more detailed simulators have appeared [1.10,1.11].

1.2.3 Conceptual Design Activities

Conceptual design includes four fundamental activities: 1) Information modeling, 2) Tradeoff analysis, 3) Partitioning, and 4) Design heritage capture.

Information Modeling - Information models are used to capture the content and meaning of the elements of a design problem. The elements which need to be represented include design requirements, constraints and budgets, technology and material descriptions, and process details. Information models are formed by identifying and classifying the objects corresponding to the technology, materials, and processes necessary for performing tradeoff analysis and design. Once objects are identified they can be converted to standard representations like EXPRESS [1.12] and compiled into databases for use in the design process. Since packaging technology involves a close relationship between design, processes, and materials it is not always sufficient to characterize materials or technologies in terms of only their measurable properties in the absence of design knowledge and process understanding. For example, knowing thermal coefficient of expansion (TCE) of a material is necessary, but the TCE differences between adjacent portions of a design may be more critical to the correct operation of the system. Design, materials, and processes interact in a non-trivial fashion which is critical to defining complex packaging systems. Thus the key issue in setting up a database is to determine how to capture, represent, and manipulate different bodies of knowledge and their interactions.

Information modeling is not strictly necessary for conceptual design to take place. Detailed tradeoff analysis can be performed without it and hardware specifications can be formulated. However, if an automated approach to tradeoff analysis is to be used (i.e., a tradeoff analysis CAD tool), then being able to describe a system and the packaging technologies associated with it in a standard form is desirable.

Tradeoff Analysis - Studies which provide comparisons of various packaging technologies contain vital information, but attempting to draw general conclusions about the applicability of one technology over another from these studies can be dangerous. Simple technology comparison studies don't provide formal methodologies to facilitate tradeoff analysis or partitioning, and therefore their conclusions are hard to apply to specific designs.

Tradeoff analysis can be addressed in either advisory or estimation-based predictive ways. Advisory methods present qualitative arguments to support or reject design decisions, while estimation-based predictive methods compute quantitative measures of system performance from which design decisions are made. Design advice explains to the designer how options compare and tracks compatibility amongst selected choices. Design advice is usually implemented using knowledge-base approaches. Besides obtaining qualitative design advice, the designer can benefit from quantitative feedback on the implications of design choices. Unfortunately, detailed simulation and physical design activities are usually too complex and time consuming to be a viable part of a broad tradeoff performance evaluation, so simplified estimation metrics are used to help make detailed tradeoff analysis practical.

Partitioning - A complex system can be partitioned in a vast variety of ways. Partitions are needed to accommodate varied design constraints, to take advantage of existing components and subsystems, to facilitate parallel development activities, to provide access, to environmentally protect, for testability, for maintainability, and to achieve reconfigurability. System partitioning is usually accomplished using highly subjective judgements. CAD tools for partitioning the functionality within chips have existed for some time. The next level in the hierarchy is the partitioning of components onto modules or boards. Module partitioning is essentially analogous to its chip-level counterpart although more interdisciplinary.

Design Heritage Capture - The rationale behind how design decisions and assumptions were made during the specification portion of the design cycle are generally lost because of poor documentation of the decision process. The lost effort becomes most apparent during the product maintenance phase when a significant portion of an engineer's time may be spent trying to recover the lost rationale that resulted in the system structure. Much of the careful deliberation and design learning that was expended on resolving key design issues is wasted, thereby increasing the overall system lifecycle costs and often contributing to longer design cycles for future products.

1.3 The Benefits of Conceptual Design

The major benefit of structured conceptual design is that the process of specifying the system to be produced is made easier because conceptual design offers a means of organizing and analyzing a collection of system implementation approaches in a rigorous and consistent manner. Given the large number of possible assembly and interconnect combinations, it is not difficult to imagine a designer hopelessly lost in analyzing and keeping track of all the interdependencies and tradeoffs. The significant impacts which conceptual design and tradeoff analysis development can provide include the following:

- Reduced design cycle time through a reduction in the number of design iterations.
- Reduced risk of inserting leading-edge technologies.
- Elimination of redesigns and/or fixes later in the design and production cycle.
- Reduced time for test and diagnosis during manufacturing and field maintenance.
- Reduced design and manufacturing cost.
- Accelerated time-to-market resulting in increase market share.

- Improved design quality.
- Improved reuse of proven designs.

Conceptual design aims to reduce a product's total time-to-market by reducing or eliminating the wasted effort associated with design modifications that are mandated to correct technology choice, testability, and manufacturing problems late in the design cycle. Tools for conceptual design will also exploit the ideas of database progressive generation and test data reuse, thus eliminating the need to regenerate well-proven technology, manufacturing, and test data and significantly speeding up development time. And, since development time has been growing at an alarming pace in recent years (many companies attribute more than half of their design cycle time to tradeoff analysis and test development), the potential impact on time-to-market can be enormous.

The objective of conceptual design is to produce designs that achieve the optimal balance in performance, cost, and manufacturing while meeting the critical needs of reliability and customer satisfaction. Conceptual design tools will enhance the quality of a design by improving the methods by which technology and test decisions are made early in the design cycle. By reducing the design time in other areas, more time is now available to apply effort in design-for-quality.

1.4 The Design Environment

The activities involved in the design and lifecycles of multichip systems include: Requirements Capture, Specification, Synthesis, Physical Design, Simulation, Manufacturing, and Customer Support. Requirements capture is the process of defining and formalizing the functionality and constraints which apply to the system. Specification is the formulation of an executable system specification from the requirements. Synthesis is the generation of a netlist and test vectors from the executable specifications. Physical design is the conversion of netlists to module and system layouts. Simulation is the detailed analysis of designed structures in support of all the other design activities. The combination of synthesis with elements of specification (supported by simulation) is sometimes referred to as a "package compiler" (or if chip design is included and multiple modules or boards are considered, "system compiler"). Manufacturing initially represents prototyping and later the production of the system for customers. Customer support encompasses all activities associated with training, maintenance, field service, etc.

Traditionally, the serial design methodology has dominated the design process, Figure 1.6. In an ideal serial methodology, each phase of the design is completed before the next is initiated. In practice, however, design iterations are always necessary, causing delays in time-

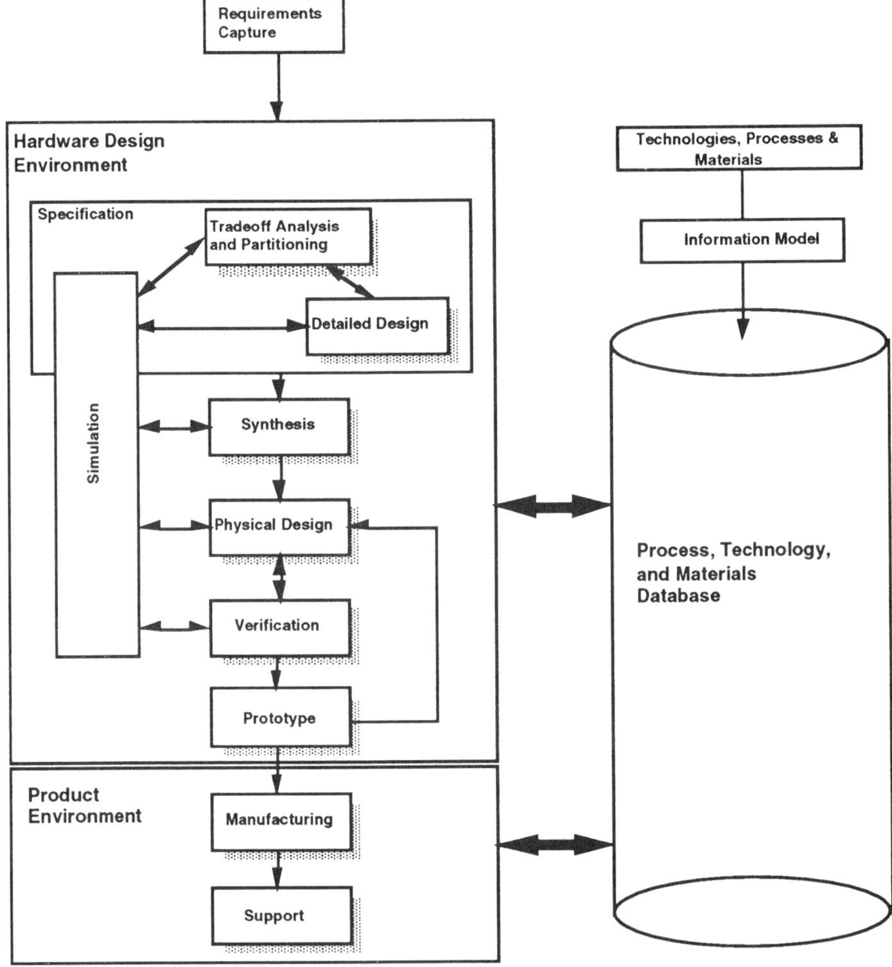

Figure 1.6. An example of a traditional serial methodology for complex electronic systems design. Ideally each activity is completed before the next is initiated.

to-market while consuming valuable resources. Traditionally, serial design methodologies relegate issues such as system testability and manufacturability to late in the design process (often after prototypes have been constructed). Whereas, the treatment of test and manufacturing issues early in the design cycle could lead to substantial savings in many cases. Serial design methodologies have evolved from a "correct by verification" sequential design approach which tends to push designers into a "depth-first" style which favors synthesis of a design before verifying that it satisfies all the design constraints.

Modern design philosophy advocates the use of concurrent design methodologies which encourage pursuing all facets of design and development in parallel. Concurrent design philosophies are built upon a "correct by design" hypothesis. One representation of a concurrent design approach for multichip systems is shown in the design/lifecycle vision in Figure 1.7. This model is referred to as a "datacentric" model,

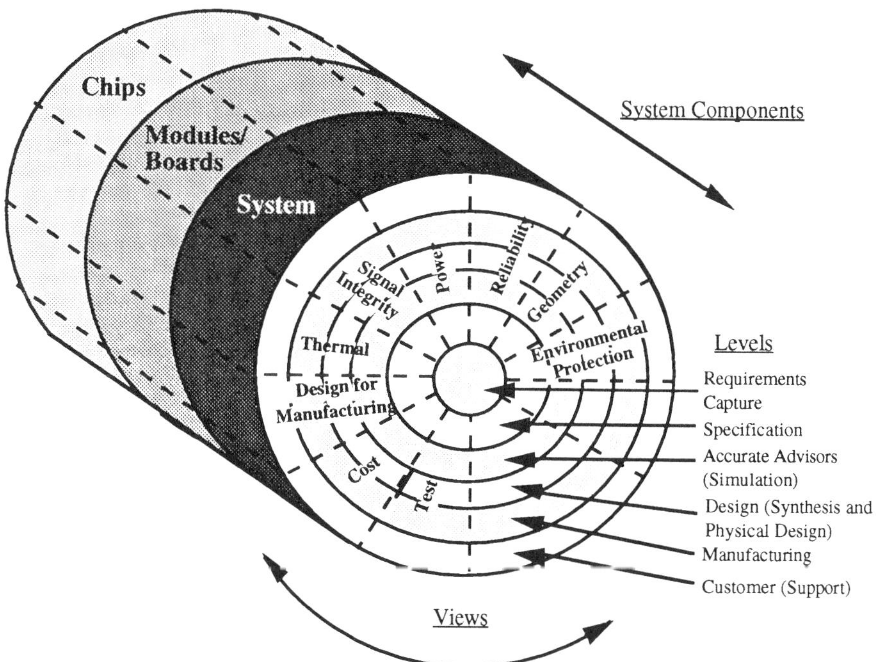

Figure 1.7. Datacentric model for the concurrent design and lifecycle of complex electronic systems, [1.13]. (© 1993 IEEE)

i.e., the designer begins at the center with very little data or information about the system (just ideas), the increasingly larger concentric rings represent increased system knowledge. The rings represent various *levels* of activity in the design, production, and lifecycles. The specification level formulates an executable system specification from the requirements and is the focus of tradeoff and partitioning activities. The concentric rings in the model are divided into sections representing various *views* which must be considered in the design/lifecycle. The topics included on the figure are important to consider, but do not represent all the factors which may have to be considered. The pie shaped elements cover all the rings, indicating that each view should be considered at all levels of design and production (i.e., design for manufacturability should be considered at the specification level in addition to its traditional treatment at the conclusion of the serial design cycle). Three different types of information flow are necessary for this concurrent design methodology. Integration is necessary between design views, between design levels, and between system components.

A concurrent methodology that contains the tradeoff analysis capabilities discussed above infers a "breadth-first" approach. The approach, explores many possible implementations prior to the synthesis of a design, thereby reducing the risk of extensive redesign later in the lifecycle.

Conceptual design that considers only a single product may not be good enough in the future. Ideally, the design environment must be able to consider a multi-product view. For example TAB bonding of a particular chip in a module may not represent the optimal cost/performance/test solution if the module is viewed as an isolated product. However, if the module is viewed as a step toward a future product, TAB bonding of the chip in today's module may be the optimum, i.e., the cost/performance of the whole product family may be optimized by beginning the insertion of the TAB process into the manufacturing line in the earlier product.

1.4.1 System Synthesis

Design synthesis is the process of producing an artifact that satisfies some high level behavioral and structural specification. Synthesis activities can be placed in perspective using a Y-chart [1.14]. The Y-chart was originally intended to demonstrate the behavior of VLSI component

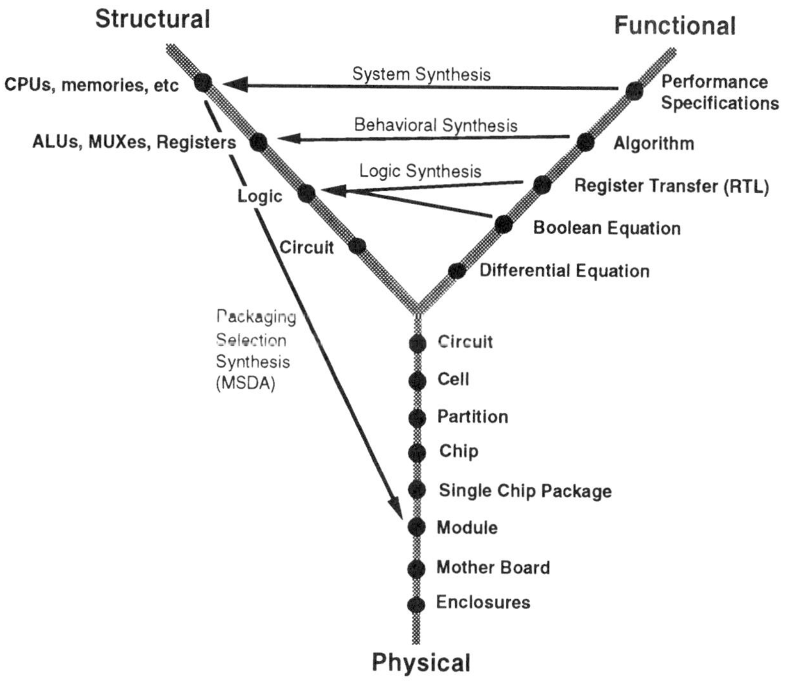

Figure 1.8. Synthesis tasks represented on a Y-chart. Common synthesis activities are plotted as mappings on the chart. One of the tools discussed later in the text (MSDA [1.13]) is included in the chart.

design tools but has been extended to illustrate system level tasks as shown in Figure 1.8 [1.15]. The axes represent three design perspectives: 1) Functional (or behavioral) - the input/output of the system; 2) Structural - the interconnection of components to realize the required functionality; and, 3) Physical - the physical manifestation of the components and interconnections used to realize the required functionality. Along each axis, multiple levels of abstraction, or levels of detail, are represented.

In general, synthesis activities map perspectives from one axis to another in a counter clockwise direction. Synthesis activities near the center of the Y-chart are generally better defined and more mature than those farther from the center. The synthesis process of interest in this book is the selection of packaging and interconnection technologies from a performance specification and a component set.

1.5 CAD Tools for Conceptual Design

Computer Aided Design (CAD) tools for both integrated circuit and packaging and interconnect design have concentrated on, and continue to concentrate in the physical design and simulation areas.

Many of the critical elements of design advisor/conceptual design technology exist today, however there have been very few attempts to pull interdisciplinary advisors together to perform concurrent synthesis and/or tradeoff analysis activities. There are two categories of existing tools which are applicable to conceptual design: 1) advisors which perform analysis or evaluation activities in one particular discipline (for one design view) and 2) design advisor managers which integrated various advisors together and manage their use.

There are a number of tools available now which could provide design advisor assistance for a particular design view at a pre-physical design level. These tools include advisors for reliability, manufacturability, cost analysis, testability, thermal analysis, and other areas. In some cases, applicable analysis modules are embedded within CAD tools performing other tasks. For example, constraint-driven routing tools may contain analysis modules which are applicable in some cases to the specification and synthesis areas. A few tools attempt to concurrently treat a subset of the interdisciplinary system synthesis/specification space using a mixture of estimation, simulation, and synthesis techniques. What sets these tools apart from constraint driven routing and extraction based physical design tools is that they are capable of operating at a pre-physical design level (i.e., without a netlist).

The estimations necessary for view specific advisors and the concurrent use of advisors are discussed at length in Section 3.3 and existing tools for tradeoff analysis are reviewed in Section 3.4.

1.6 Organization of This Book

This book is organized into two parts. The first part (Chapter 2) discusses the representation of information and details how design elements are described, and design constraints are captured and represented so that design tools can make use of them. Chapter 2 also discusses capturing the design decision process. The second portion (Chapters 3, 4, and 5) concentrates on the tradeoff analysis and partitioning portion of conceptual design. The methodology necessary for computing the performance metrics at the heart of tradeoff analysis are derived and discussed in detail in Chapter 3. Chapter 4 treats the multichip partitioning problem, and Chapter 5 presents examples of tradeoff analyses performed for different multichip systems.

1.7 References

[1.1] *Microelectronic Packaging Handbook*, editors R. R. Tummala and E. J. Rymaszewski, Van Norstrand Reinhold, 1989.

[1.2] D. P. Seraphim, C. Y. Li, and R. Lasky, *Principles of Electronic Packaging, Design and Material Science*, McGraw-Hill, 1989.

[1.3] *Multichip Module Technologies and Alternatives - The Basics*, editors D. A. Doane and P. Franzon, Van Nostrand Reinhold, 1993.

[1.4] C. F. Coombs, *Printed Circuit Handbook*, McGraw-Hill, 1988.

[1.5] A. J. Prasad, *Surface Mount Technology Principles and Practice*, Van Norstrand Reinhold, 1989.

[1.6] A. Kozak, C. Harris, K. Hubbard, J. Rostworowski, and I. Verigin, “MCMs in Telecommunications Designs,” *Surface Mount Technology*, vol. 5, no. 3, pp. 46-49, March, 1991.

[1.7] R. O. Carlson, C. W. Eichelberger, R. J. Wojnarowski, L. M. Levinson, and J. E. Kohl, “A High-Density Copper/Polyimide Overlay Interconnection,” *Proceedings of the Eighth International Electronics Packaging Conference*, pp. 793-804, 1988.

[1.8] *Thin Film Multichip Modules*, editors G. Messner, I. Turlik, J. W. Balde, and P. E. Garrow, ISHM Press, 1992.

[1.9] J. W. Rozenblit, J. L. Prince, and O. A. Palusinski, "Towards a VLSI Packaging Design and Support Environment (PDSE) Concepts and Implementation," *Proceedings of the International Conference on Computer Design*, pp. 443-448, 1990.

[1.10] G. Choksi, B. K. Bhattacharyya, S. Stys, and B. Natarajan, "Computer Aided Electrical Modeling of VLSI Packages," *Proceedings of the 40th Electronic Components and Technology Conference*, pp. 169-172, 1990.

[1.11] P. A. Sandborn and H. Hashemi, "A Design Advisor and Model Building Tool for the Analysis of Switching Noise is Multichip Modules," *Proceedings of the International Symposium on Microelectronics (ISHM)*, pp. 652-657, 1990.

[1.12] "ISO, EXPRESS Language Reference Manual, External Representation of Product Definition Data," *ISO TC184/SC4/WG5 N14*, 1991.

[1.13] P. A. Sandborn, K. Drake, and R. Ghosh, "Computer Aided Conceptual Design of Multichip Systems," *Proceedings of the Custom Integrated Circuits Conference*, pp. 29.4.1-29.4.4, 1993.

[1.14] D. D. Gajski, and R. H. Kuhn, "Guest Editors' Introduction: New VLSI Tools," *IEEE Computer*, vol. 16, no. 12, pp. 11-14, December, 1983.

[1.15] W. P. Birmingham, A. P. Gupta, and D. P. Siewiorek, *Automating the Design of Computer Systems, The MICON Project*, Jones and Bartlett Publishers, 1992.

Chapter 2

Information Modeling and Representation

2.1 Introduction

Information modeling is an activity which abstracts and represents the important characteristics of a complex system in a form which makes those characteristics understandable and accessible to a variety of people and computer tools.

As multichip systems grow in complexity, the information contained in them originates from a large variety of sources and touches upon many different areas of expertise. It is becoming difficult for one person, or even a team of people, to know or understand all the necessary information which describes a multichip system. In the past, documentation systems were used to capture and archive a design and its information content. Today, as we stand on the eve of information highway systems and enterprise integration networks, it is becoming mandatory, from the point of view of both suppliers and customers, to have a universally accepted means of information representation that makes the design, manufacturing, and marketing tasks agile, and to optimize costs and performance through fast and accurate information access.

Consider the following situation: After several months of work, a multichip module design has been completed and is ready to begin manufacturing. At the last moment, news is received that one of the chip suppliers has raised the price of its product so that now a different supplier is more attractive. This situation requires fast and reliable electronic access to the physical descriptions of the new chips with data on availability, thermal, electrical, power dissipation characteristics, and assembly requirements. If the correct information environment were available, changing suppliers could be achieved in an orderly, timely fashion, and at a cost advantage over the alternative. Unfortunately, a complete and accurate electronic information system is not generally available. Instead, the required data is normally obtained from a variety of sources, each of which is presented in different form, and a significant portion of which does not reflect the latest information. In this example, if one needed to quickly obtain the physical description of the new chips, hardcopy images from the supplier would be the most likely format. In order for physical design tools to understand them, they must be reentered into the design (with a corresponding risk of error). For example, the functional pin conventions may vary between suppliers and the person entering the new data may forget, or incorrectly update the pin labels. Such seemingly trivial errors are common, and very difficult to detect and costly once they are introduced. In addition, the situation we are exemplifying brings up another important point. Designs are dynamic objects. They change in time because the constraints and available information change. In our example, the design had been completed with a specific set of chips. Now that the market conditions dictate a switch in a supplier, a new design emerges. The reasons for the change are clear at the time it occurs, but unless the reasons for the change are captured they may be lost.

The problem which this example illustrates is one of information management. The critical issues in information management are:

- Information representation
- Information content
- Information dynamics
- Information availability and retrieval
- Information version control

Information representation ensures that the information obtained is in a form that is understandable to the designers and their CAD tools. Information content refers to the meaning of the information. Information dynamics refers to the evolution of information content and the capture of the underlying design decisions. Information availability and retrieval refer to the ease with which the necessary information can be found and obtained. Information version control ensures that the information obtained is the most up-to-date version that exists.

This chapter focuses on the issues of information representation, content, and dynamics. Progress toward the solution of these three issues provides the necessary infrastructure to manage the other two.

A practical consideration in multichip system design is the determination of the relevant information that needs to be stored and shared. The problem is, how to communicate the necessary information in a form that is clear, unambiguous, and to the point. Now that computers (or information processors) are in widespread use, the ability to store and retrieve vast amounts of information is greatly enhanced. However, unless we can make sense out of what the computer retrieves, mountains of data will not help.

Today, computers and their electronic linking via high bandwidth communication networks, are solving much of the problem of having access to information. It is now necessary to filter nonessential information from that which is really relevant.

As exemplified above, one of the major problems faced in the electronics design industry is the need to communicate, efficiently and accurately, critical design data. At the same time, it has become necessary to control access to information considered proprietary. Even as recently as the early 1980s, many software and hardware vendors encoded and stored design information in forms they considered proprietary, with the purpose of protecting their market and keeping their customers captive. However, because of the widespread use of these languages and formats among users of the most popular commercial electronic aids, some of these languages gradually evolved into de facto "standards" of communication and information exchange. This initially took place among users of those proprietary systems and later among other designers that were not within the set of captive users.

The next section addresses how information is modeled in order to represent it in a general form. The information modeling section is followed by a section describing the main parts of the STEP (STandard for Exchange of Product data) standard, which embodies an international thrust for standardization based on the main concepts behind information modeling. After that, we present and analyze a specific information model, suitable for a portion of specification activities and physical multichip system design. Finally, the concepts of information modeling and design heritage capture are linked, with an introductory discussion of Issue-Based Information Systems (IBIS).

2.2 Modeling Information

Data, numbers for example, can be stored in a disk file. The risk is that sooner or later the subject of the file will be forgotten, and soon thereafter the information content is totally lost.

If the information consists of a variety of data types, it becomes more difficult to manage and upkeep it in a disk file. Searching for data of certain type, and worse yet, associating different types of data becomes, in general, a lengthy process, and there is no unifying principle as to how the data is organized, nor an easy way to communicate to other people what the file contains.

There are two issues that need to be addressed to properly deal with information. Its content and its form, or semantics and syntax. This is subject of information modeling. Information modeling, when allied with modern software methods, provides powerful solutions to many problems, including the ones mentioned above.

Semantics refers to the content or meaning of the information that is being modeled and syntax refers to the construction through which one conveys the semantic content. The content or meaning is usually expressed through a Schema, or diagrammatic-like description, as illustrated in Figure 2.1.

To introduce information modeling, consider a simple example. Assume we have data that represents straight lines of metal intercon-

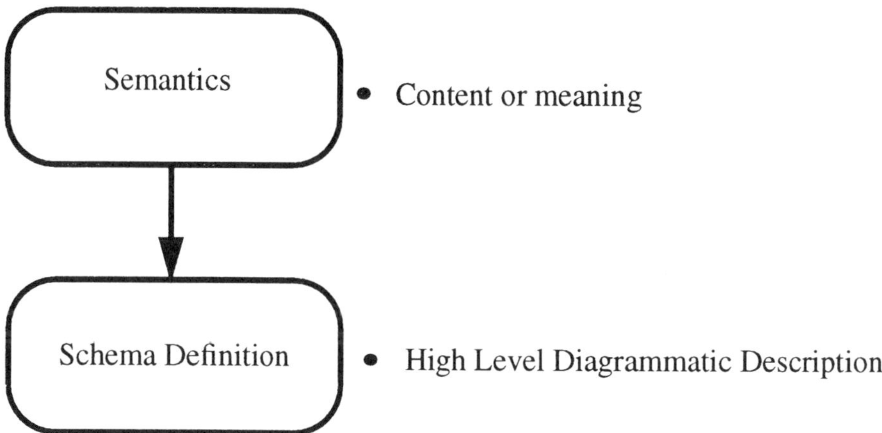

Figure 2.1. Meaning is conveyed through Schemas.

nect. The metal interconnect has geometric characteristics. If all lines lie in parallel planes, then the characteristics of interest are their length and width, expressed in some system of units. We will also be concerned with knowing in which of the several parallel planes a particular line of interconnect lies. It will also be important to know the thickness of the straight interconnect line in the direction perpendicular to the plane in which it lies. It may be important to know the dielectric properties of the medium surrounding the line and the material composition of the line itself.

To begin with, since we are going to be talking about straight lines of interconnect, let us agree to call any of these lines a "path". We could have selected any other name for the line, calling it a "line" would have been acceptable also. The important thing is we are reserving a particular name for the object or entity[1] being modeled and agreeing to use it consistently. This may seem to be a trivial issue, but it is an important first step in the information modeling process.

In order to model the information which describes the path entity, we need to start with the list of properties that describe it. A real, manufactured interconnect line is a much more complex thing than the list

1. An entity is defined as an object, having an independent existence, that represents a basic concept.

of properties we are focusing on in our simple description. For example, one physical attribute about the real interconnect line is its temperature, which in turn could be defined as a continuous physical field defined over the three dimensions where the physical interconnect exists, plus time. This level of description is far beyond our ability to represent with the simple information model we are constructing here. Instead, we will simply model the information we have about the interconnect line (and accept the fact that our model is incomplete). The fact that the model is incomplete may not be troublesome since the detailed description of the temperature field distribution may not be relevant to our design aims.

Begin by writing down a list of the information to be represented:

Coordinates of starting point
Coordinates of ending point
Width
Height
Plane designator
Properties of embedding medium
Line composition

According to the list above, we need additional concepts or entities, for example, "points". In general, defining information about complex entities requires defining information about other simpler entities. However, at the bottom of everything there is a set of primitive or basic data types. The usual set of primitive data types are things that computers can easily deal with, like numbers, which could be integer or real, and strings of characters which can convey information content in some language or convention.

In order to describe the "point" entity formally, the following description could be used:

```
Entity point;
x: REAL;
y: REAL;
End_entity;
```

This simply states that we are defining the information content of an entity we call point. It consists of two real numbers, which are referred to as x and y (more generality could be added by including z as well). It is a convention to write basic, fundamental primitive types in capital letters (i.e., REAL). This tells us at a glance that we have reached the atomic level of information, there is no substructure after that. It is also a convention to write the semicolon at the end of every statement, and the colon after the information items or entity attributes (i.e, x and y).

Now, formalizing the path entity:

```
Entity path;
begin: point;
end: point;
width: REAL;
height: REAL;
layer: INTEGER;
mediaProperties: ListOfMediumProperties;
lineComposition: crossSection;
End_entity;
```

An obvious but important property of the model above is that it is not unique. We have the choice of order of the entities in the definition, we have the choice of names of the entities, and we have the choice as to how to model certain entities. For example, lineComposition. Not only is it our choice of name, but we have decided to temporarily postpone deciding how to describe it until we define what the crossSection entity is. Why call it crossSection? The name conveys the idea that metal lines may not be homogeneous. We may be representing a metal line that has a core of copper but it is clad with nickel, for example.

The non-uniqueness of information models, and of course, the non-uniqueness in how the information is expressed is what motivates a quest for a standard or a set of them, i.e., something that everybody will agree to and use. Defining such a standard is extremely difficult. In general, people do not know in advance all the information needed for any given application. Therefore, the modern attitude among standards experts is that there should be freedom with regard to what information to store for an application, but there must be agreement with regard to how we will express the high-level description (schema), so that it can be easily ported and exchanged among people. In addition, the way in which the information model is built will help convey the meaning of the information. This brings us back to the issues of semantics and syntax which we introduced above. Semantics depends on the application, and syntax should be general and able to convey the semantics.

This concept may be expressed through a diagram. The formal definition of the path entity is part of a high-level description of the information we want to model. The full high-level description that defines the relationships among all entities is a Schema, shown in Figure 2.2. In general, an entity depends on other entities, so one can portray the situation as in Figure 2.3.

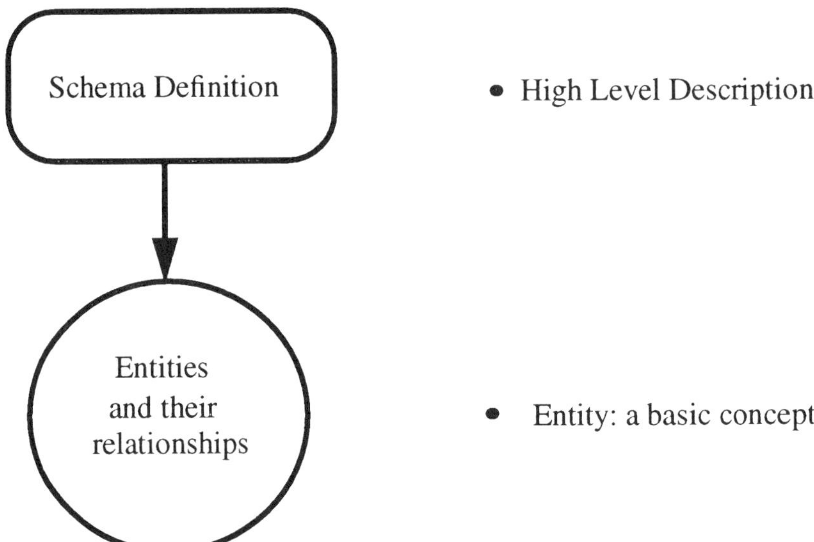

Figure 2.2. A schema describes entities and how they are interrelated.

The simple example above, which attempts formalization of the information on straight interconnect lines is written in a standard information modeling language called EXPRESS. This language has been defined by the International Standards Organization (ISO). Schemas are described through the standard language, EXPRESS, Figure 2.4.

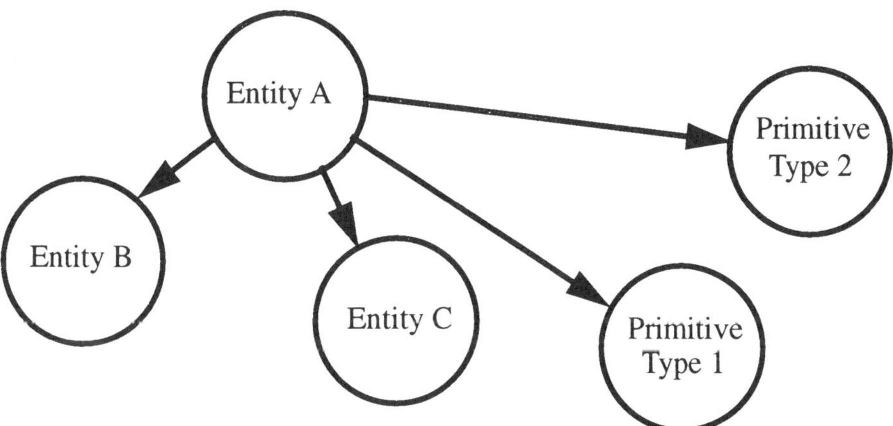

Figure 2.3. An entity is defined through its structural dependence on other entities.

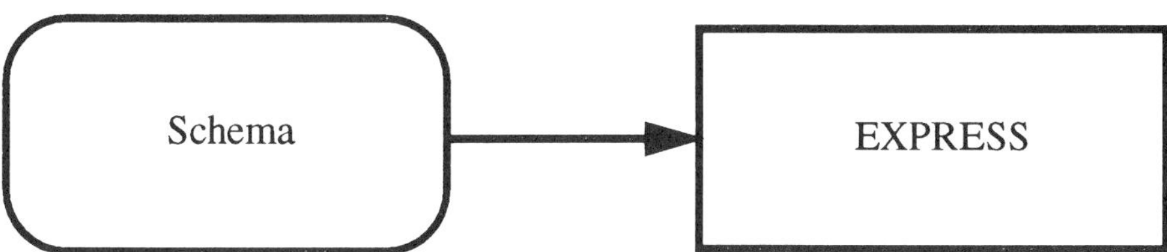

Figure 2.4. A schema is defined through a standard language, EXPRESS.

In practice, once an information model has been defined through EXPRESS, one would like to use a computer language to implement and manipulate the information base. There are advantages to using the so-called object-oriented languages, such as C++. The normal path is to associate an object definition, or class, with every entity in the information model, as indicated in Figure 2.5.

In a real implementation of the information model, a class is instantiated as many times as necessary. In our example, we would instantiate the path entity as many times as there are interconnect lines. Each of those instances will contain the individual data for the corresponding interconnect line. This situation is visualized through Figure 2.6.

The whole process of information modeling and implementation is summarized by Figure 2.7.

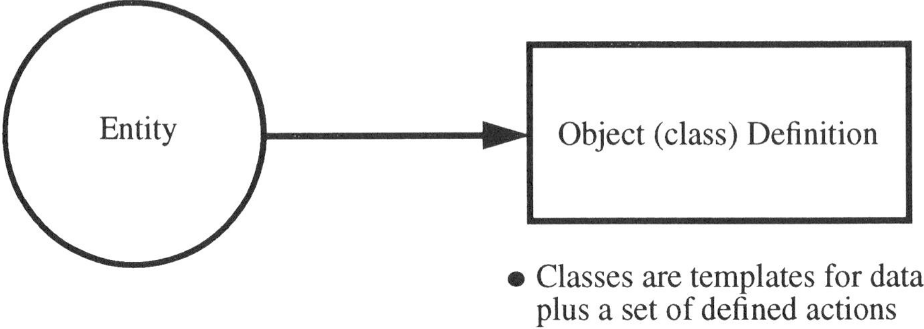

Figure 2.5. Entities are mapped into classes.

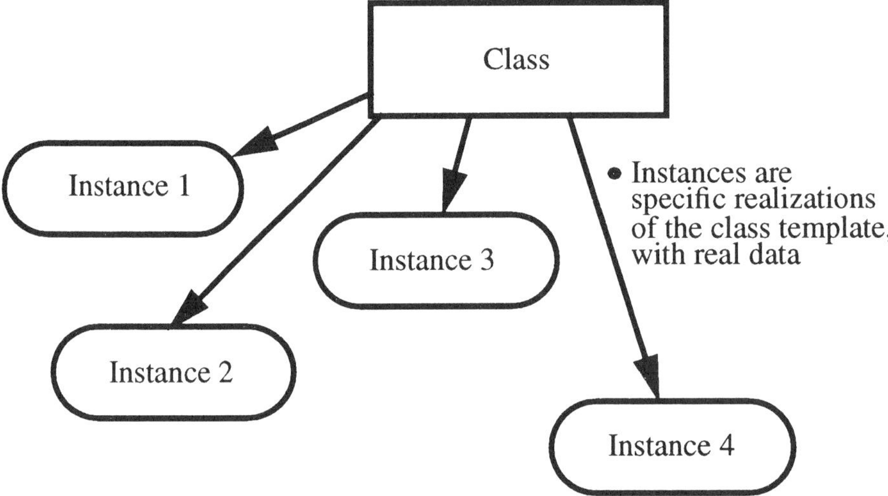

Figure 2.6. Object instances are individual copies of the class template filled out with data.

2.3 The Need for STEP (STandard for Exchange of Product data)

A variant to the problem discussed in the introduction to this chapter is the following: when a manufacturer wants to introduce a new part or product into the market, they are often faced with a problem. Even if the product is superior to what is already available, there is a reluctance to accepting it because there is no common and complete way to describe the part. In the minds of customers, there are nagging ques-

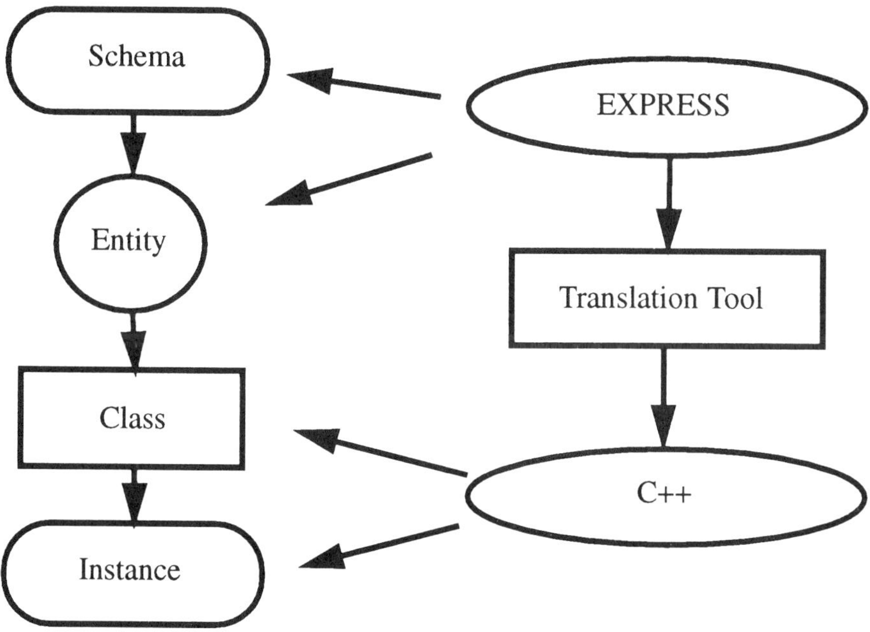

Figure 2.7. Information modeling and its implementation path.

tions: is it as efficient as what is presently used? There is also often a lack of understanding of critical quality, cost, and technical specifications. Will it fit into the integrated product or will its introduction create new and unforeseen compatibility problems? These and many other questions trouble potential customers and increase the time to market and expense in the introduction of the new product. Now, assume that there exists a way to characterize the product through sufficient measurements, and that such measurements are available for the product in question. Moreover, let us accept that there is a way to exchange this data in a form that everyone understands. Then our hypothetical manufacturer can anticipate faster introduction and acceptance of the new part. The transfer of information can in principle be bidirectional, so that the systems integrator can send back to the supplier information about discrepancies which are found, and improved communication makes it possible to accelerate the time-to-market.

This problem and many others are being addressed by the international development of STEP [2.1]. The full standard consists of several parts, which are briefly described in the Appendix to this chapter. One of the parts deals with the EXPRESS language (introduced in Figure 2.4). The Appendix to this chapter provides an introduction for readers who are not familiar with STEP and/or the EXPRESS language. A complete description of STEP can be found in the documents published by ISO TC184/SC4.

One of the reasons why industry is embracing the STEP standard is that traditional database implementations suffer from limitations on the information that they can efficiently store. In general, traditional database approaches require a fixed-length record information representation and/or tables. The information models one defines with EXPRESS are of a more general nature than those which can be mapped into tables and/or fixed-length records. Many product characteristics, and especially those that are of interest in engineering design are ill-suited for representation using tables or fixed-length records. By contrast, object-oriented representations appear to be better suited, and EXPRESS language constructs can be easily mapped into any of a number of object programming languages. The example in the following section illustrates how EXPRESS appropriately represents information that is useful in design applications.

2.4 A Conceptual Design/Layout Schema Example

We now introduce a practical EXPRESS schema. The explanatory graphics beside the EXPRESS statements are not part of the schema, they are added for clarification purposes.

The schema is applicable to both conceptual level tradeoff activities and physical design. Note that early tradeoff analysis does not necessarily make use of all the information which the schema contains.

```
(*****************************************************)
(* *)
(* Copyright 1993, Microelectronics and Computer Technology *)
(* Corporation *)
(* All rights reserved *)
(* *)
(*****************************************************)
SCHEMA Twodimgeometry;

ENTITY Characteristic;
name: STRING;
value: STRING;
END_ENTITY;
```

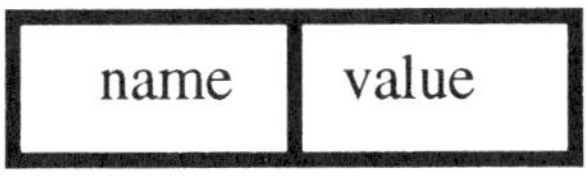

name | value

```
ENTITY Layer
SUBTYPE OF (GeometricObj);
lyr: INTEGER;
END_ENTITY;
```

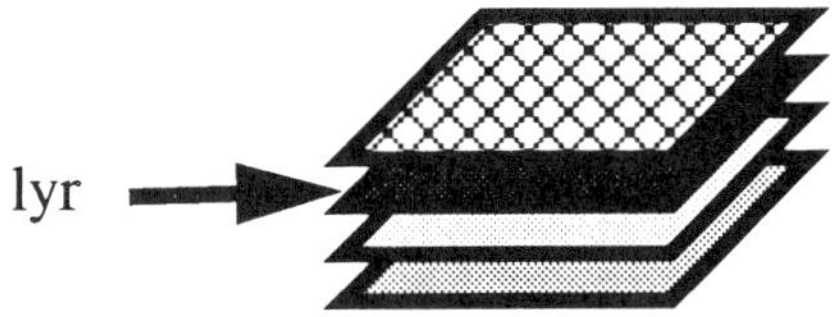

lyr

```
ENTITY Point
SUBTYPE OF (GeometricObj);
x: REAL;
y: REAL;
END_ENTITY;
```

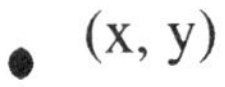

(x, y)

```
ENTITY BoundingBox
SUBTYPE OF (GeometricObj);
lowleft: Point;
upright: Point;
END_ENTITY;
```

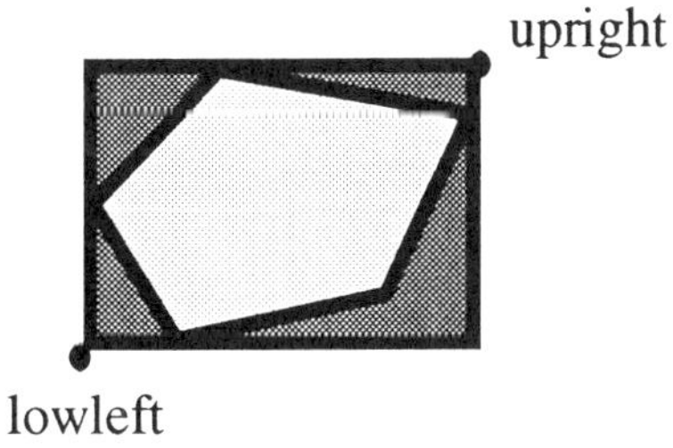

upright

lowleft

```
ENTITY Slimline
SUBTYPE OF (GeometricObj);
lyr: Layer;
nPath: INTEGER;
path: LIST OF Point;
END_ENTITY;
```

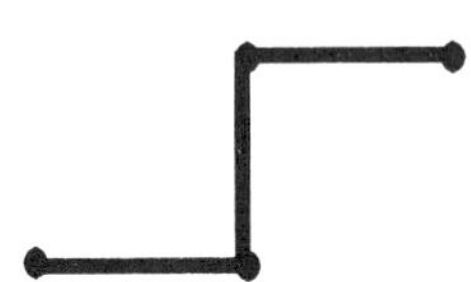

```
ENTITY Wideline
SUBTYPE OF (GeometricObj);
beginExt: REAL;
endExt: REAL;
lyr: Layer;
netNum: INTEGER;
nPath: INTEGER;
path: LIST OF Point;
pathShape: STRING;
width: REAL;
END_ENTITY;
```

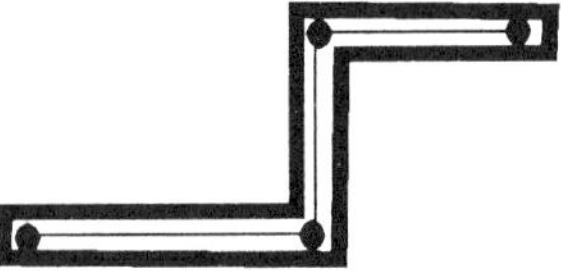

```
ENTITY Box
SUBTYPE OF (GeometricObj);
bBox: BoundingBox;
lyr: Layer;
END_ENTITY;
```

```
ENTITY Boundary
SUBTYPE OF (GeometricObj);
lyr: Layer;
nPath: INTEGER;
path: LIST OF Point;
END_ENTITY;
```

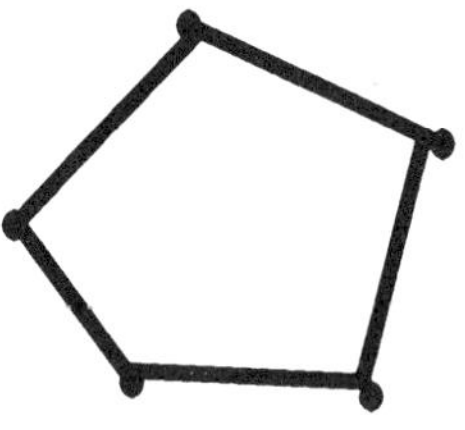

```
ENTITY CircularArc
SUBTYPE OF (GeometricObj);
center: Point;
begin: Point;
end: Point;
lyr: Layer;
END_ENTITY;
```

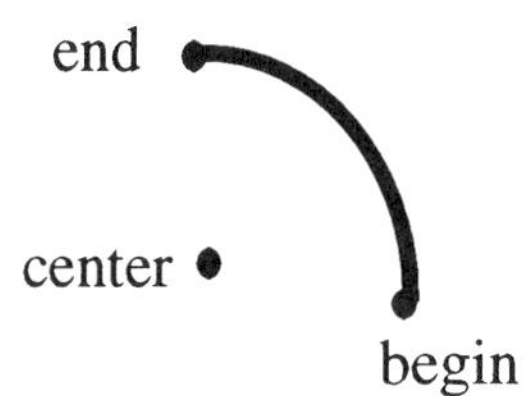


```
ENTITY Textstring
SUBTYPE OF (GeometricObj);
draftingP: BOOLEAN;
font: STRING;
height: REAL;
justify: STRING;
labelType: STRING;
lyr: Layer;
orient: STRING;
angle: REAL;
theLabel: STRING;
xy: Point;
END_ENTITY;
```

Some_Text

xy

```
ENTITY Structure
SUBTYPE OF (GeometricObj);
blockName: STRING;
structType: STRING;
objList: LIST OF GeometricObj;
END_ENTITY;
```

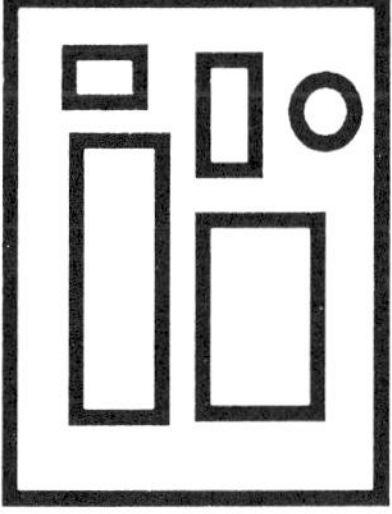

```
ENTITY Occurrence
SUBTYPE OF (GeometricObj);
master: STRING;
blockRef: STRING;
orient: STRING;
xy: Point;
END_ENTITY;
```

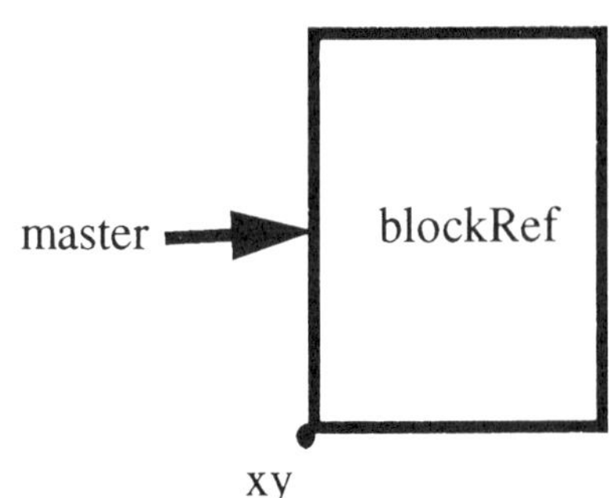


```
ENTITY Array
SUBTYPE OF (GeometricObj);
columns: INTEGER;
rows: INTEGER;
master: STRING;
uX: REAL;
uY: REAL;
xy: Point;
orient: STRING;
END_ENTITY;
```

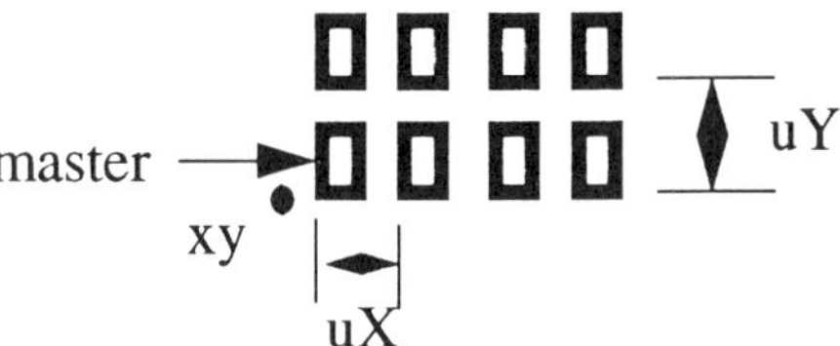


```
ENTITY ElementPin
SUBTYPE OF (GeometricObj);
pinName: STRING;
pinNum: INTEGER;
padRef: STRING;
pkgRef: STRING;
END_ENTITY;
```

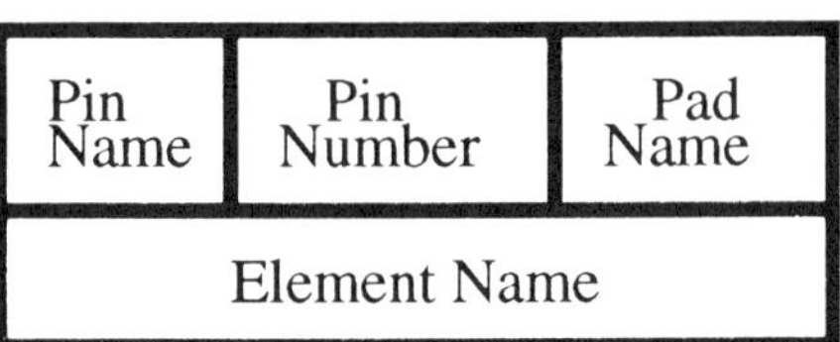


```
ENTITY Signal
SUBTYPE OF (GeometricObj);
sigName: STRING;
sigNum: INTEGER;
elPinList: LIST OF ElementPin;
END_ENTITY;
```

List of ElementPin	
Signal Name	Signal Number

```
ENTITY GeometricObj
SUPERTYPE OF (ONEOF (Point, Layer, CircularArc, Slimline,
Wideline, Boundary, Box, Textstring, Structure, Occurrence,
ElementPin, Signal));
selfIdent: STRING;
props: LIST OF Characteristic;
END_ENTITY;
```

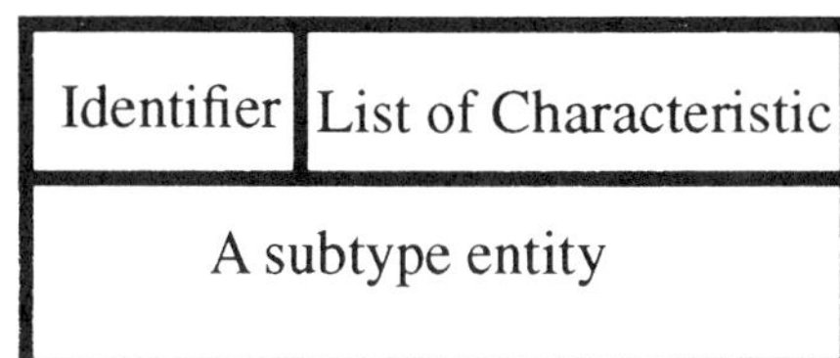


```
END_SCHEMA;

(**************************************************)
```

The schema above models information suitable for describing two-dimensional drawings which contain electrical connections. Everything in the schema, with the exception of the entity *Characteristic*, is considered to be a geometric object, referred to as *GeometricObj*. The rationale behind this, is that the named entity attributes (in each entity definition) are qualities of a general nature. The individual instances of an entity may possess local properties which are then attached to it in its guise as *GeometricObj*. One can turn this argument around and say that any entity will possess a list of characteristics, and from the formal point of view, it is more economical and elegant to describe the fact that everything has common characteristics by defining the supertype *GeometricObj*, and making every subtype inherit those general characteristics. This negates the need to define a LIST OF Attribute attribute for each and every entity.

Most entities in the schema are of a geometric nature[2]. In addition to the geometric entities there are two electrical connectivity entities, *ElementPin* and *Signal*.

---

2. We have chosen not to refer to the entities in Part 42 of ISO 10303 because our schema is short and self contained. Our simple schema attempts to describe entities in two dimensions which limits the need for the more general entity definitions in the standard.

From the designer's point of view, the most complex entity is the *Structure*. An individual *Structure* is able to have as many attributes as needed. What makes a *Structure* a top object in the information hierarchy is its attribute objList, which states that it can contain any number of other *GeometricObj's*. A rule, which is not implied in the schema as we have written it, is that we will not allow a *Structure* to contain another *Structure* in its objList. EXPRESS is a language that can incorporate constraints through the use of constructs called Rules, combined with the ability to define functions and procedures. In this text, we will not cover these more advanced capabilities of the language, because in the practical implementation of the schema, presently available programming tools do not enforce rules and constraints easily.

In practice, one defines a *Structure* by constructing it out of other *GeometricObj's*, and generates the objList, checking that as an object is added, it does not happen to be a *Structure*. If an attempt to add a *Structure* occurs, it is signaled as an error.

A *Structure* can use the contents of another *Structure* through two entity types: *Occurrence* or *Array*. The *Occurrence* makes the contents of another *Structure* (here again, we have the constraint that a *Structure* must not have an occurrence of itself) appear at a point xy with a certain orientation (combination of rotation and mirroring or reflection). An *Occurrence* entity has an attribute defined as master, the name of the *Structure* being instantiated, and a unique name (in this schema known as blockRef) to distinguish it from other occurrences of the same master that may appear in the *Structure's* objList.

An *Array* is a simple array of occurrences. As such and as defined in the schema, *Array* is not useful when the array needs to have electric connectivity associated with it, because we do not have a way to distinguish or address the individual elements. Instead, an *Array* is used as an abbreviation for an array of drawn occurrences. Whether or not a *Structure* should appear in an *Array* depends on the value of its attribute structType. Once again, there is the constraint here that a *Structure* containing an *Array* should not reference itself.

The issue of constraint satisfaction we have been referring to is a complex one. In our case, and looking at the *Occurrence* or *Array* cases, a problem which arises is the potential for cycles, because an instantiated *Structure* may, deep within its own hierarchy, instantiate the very *Structure* we may be attempting to define. The only solution to this dilemma is to completely check the descendants in every *Structure's* hierarchy and make sure one ends up with a tree graph. This can be effectively done only after the full set of *Structures* which comprise a design has been built.

With the exception of *Signal* and *ElementPin*, the connectivity entities, we will not discuss in detail the remaining entities because they are fairly straightforward and self-explanatory. It is obvious that one can think of more constraints that could be included in the schema. For example, *Slimlines*, *Widelines*, *Boundaries* need to be non-reentrant, or non-self-crossing. Any geometric entity could be subject to a constraint that demands it to have a bounding box smaller than the maximum bounding box physically allowed by a particular design application, and so on. In practice, the management of these constraints is left to the software that builds the data. If data is already built and presented to us, the software that uses it needs to verify constraint satisfaction. A better schema than the one presented will have all constraints defined within it and will have programming tool support to guarantee that actual entity realization will have the state of every constraint checked. EXPRESS defines three states for a rule: asserted (or satisfied), violated, or unknown. If a rule is violated, then the information base does not conform to the EXPRESS model.

An *ElementPin* is a collection of information that tells us that an *Occurrence*, with its unique element name (blockRef in this schema) has an electrical port, referred to here as a pin, and described by a number and/or a name. Associated with this port there may be a specific pad reference, which would be the name of the *Structure* that physically implements the port, and which is to be found somewhere in the hierarchy of the *Occurrence* master.

The *Signal* entity contains the net name and/or number and a list of all the electric ports that it connects, in the form of a list of *ElementPin* entities. Because both *Signal* and *ElementPin* entities are declared as *GeometricObj's*, they can coexist with the geometric cntities they refer to in the same *Structure*.

The schema we have been discussing is similar to one used for transferring information between CAD systems [2.2]. The basic mechanism is provided by a persistent object representation of the EXPRESS entities, implemented through C++ classes, together with a suite of persistent object manipulation methods which allow conversion of the stored information into customized representations that different CAD systems accept [2.3]. This approach allows successful concurrent, parallel design of all the various components of a complex multichip system using a set of heterogenous CAD tools, in a fraction of the usual time, and allows optimal utilization of available design resources.

2.5 Capturing Design Heritage Information (Information Dynamics)

When dealing with complex design problems at an early design stage, there are many possible choices (tradeoffs) that can be made with regard to any single design issue. As each issue is considered and decisions are made, the final solution to the design problem takes shape. As the design evolution progresses, problems that occur due to earlier decisions become apparent, but the reasons upon which those earlier decisions were based are sometimes forgotten. The end result is that there are reversals of position in midstream, forcing unnecessary reconstruction of earlier tradeoffs. Ideally one would like to have a record of what decisions had to be made, what the possible choices were, and what the arguments for and against each of those choices were.

One approach to solving this problem is provided by Issue-Based Information Systems (IBISs) [2.4], [2.5]. The idea behind IBISs is that complex design problems are solved by casting them into issues to be resolved through conversations, or deliberations. These conversations take place among different members of a design team and expert CAD tools. Every issue proposed for discussion generates a number of positions, and every position generates a number of arguments, some in its favor, others against.

An issue is something that can be cast as a question of a controversial nature, so that there are several, usually conflicting positions that represent possible answers to the question. The arguments in favor or against each position help the design team choose which position to select in order to settle the issue.

A given issue may, in the process of discussion, open up new issues. Each of those will in turn generate positions, and each of those their own pro and con arguments. One can select a position in answer to an issue, but the sub-issues generated must be analyzed as well. The problems alluded to in the first paragraph of this section arise because positions (i.e., choices in response to design problems) are not fully investigated before a choice is made (a position selected). Thus, if issues appear that seem to invalidate an earlier decision, it can be of great help to have the history of what transpired previously, and the alternate positions with their pro and con arguments.

The information modeling discussed in the first half of this chapter, can be coupled with the IBIS approach. A very simple EXPRESS model for an IBIS is presented below.

```
(*****************************************************)

SCHEMA IBIS;

ENTITY Issue;
theIssue: STRING;
Positions: LIST OF Position;
Subissues: LIST OF Issue;
status: STRING;
END_ENTITY;

ENTITY Position;
thePosition: STRING;
theIssue: Issue;
ArgumentsFor: LIST OF Argument;
ArgumentsAgainst: LIST OF Argument;
subIssuesRaised: LIST of Issue;
status: STRING;
END_ENTITY;
```

```
ENTITY Argument;
theArgument: STRING;
thePosition: Position;
the Issue: Issue;
status: STRING;
END_ENTITY;

END_SCHEMA;

(*****************************************************)
```

As discussed earlier in this chapter, information models such as the one above can be put to practical use by implementing them via object-oriented programming. Tools are available that can map the schema above into C++ classes, one for each entity. Supporting software (object handling methods) are then created so that as a design problem is considered, the issues (and possible sub-issues) raised can be stored as persistent objects. Positions, i.e., alternate choices, can also be captured (stored as persistent objects), and in the same way, all the arguments (reasons) for and against each choice. The whole chain of events leading to a design decision can thus be preserved and the reasons for a particular choice, as well as alternate possibilities kept available if the need to reconsider them arises.

To illustrate the IBIS method, consider the following exercise. A system design house is constructing a multichip module. The design team must decide whether to bond a particular chip using wirebonding, TAB, or flip chip:

> ISSUE A: What bonding method should be used?
> POSITION 1: Wirebonding should be used.
> ARGUMENT 1a (in favor): Well developed infrastructure exists.
> ARGUMENT 1b (in favor): Low tooling cost.
> ARGUMENT 1c (opposed): High inductance (possible switching noise concerns).

ARGUMENT 1d (opposed): Bare die test and burn-in difficult.
POSITION 2: Tape Automated Bonding (TAB) should be used.
ARGUMENT 2a (in favor): Test and burn-in can be done on tape.
ARGUMENT 2b (in favor): Controlled impedance bond.
ARGUMENT 2c (opposed): High tooling cost.
ARGUMENT 2d (opposed): High inductance (possible switching noise concerns).
ARGUMENT 2e (opposed): Lack of infrastructure.

POSITION 3: Metallurgical flip chip bonding should be used.
ARGUMENT 3a (in favor): Low inductance.
ARGUMENT 3b (in favor): Smaller area.
ARGUMENT 3c (opposed): Bare die test and burn-in difficult.

SUBISSUE A.1: Does necessary infrastructure exist?
POSITION A.1a: Yes, wirebonding infrastructure is mature.
ARGUMENT A.1a (in favor): Over 1000 wirebond suppliers exist in the world.
ARGUMENT A.2a: ...

POSITION A.1b: No, based on the necessary production volume there are no domestic TAB suppliers who can guarantee delivery on time.
ARGUMENT A.1b: ...

SUBISSUE A.2: Is there a switching noise problem?
POSITION A.2a: No, the present design has rise time greater than 2 ns.
ARGUMENT A.2a: ...

POSITION A.2b: Yes, the planned extension to this product will use drivers with rise times less than 1 ns.
ARGUMENT A.2b: ...

IBISs capture and store the information in a way that allows future designers to retrace the decision making process and understand the basis for critical design decisions.

APPENDIX - An Introduction to STEP and EXPRESS

STEP. In order to promote standardization and provide solutions to the general problems facing information exchange, in 1984, the International Standards Organization (ISO) established a subcommittee (SC) of Technical Committee 184 (TC184, Technical Committee on Industrial Automation Systems and Integrations). The official designation of the subcommittee is: ISO TC184/SC4. The mission of this body is to develop international standards to represent and exchange product information, the standard is known as STEP. When STEP meets its goals, it will provide a solution to the product data exchange problem, and by extension, to many of the analogous problems faced in the engineering of products. The standard is being developed at this time and is known as ISO 10303, Industrial Automation Languages: Product Information Representation and Exchange [2.6]. ISO 10303 is divided into many parts that fall into four types: Tools, Integrated Generic Resources, Integrated Application Resources, and Application Protocols.

The tools portion of ISO 10303 consists of: Part 1 - an introduction and description of STEP, Part 11 - definition of the EXPRESS language used to develop STEP information models, Part 21 - specification of the exchange file format, and Part 31 - introduction to STEP conformance testing.

The integrated generic resources portion of ISO 10303 consists of: Part 41 - fundamental principles of how to describe products and the resources referring to them, such as approval, person and organization, measurement units and so on. Part 42 defines geometric and topological representation, Part 43 refers to representing concepts and their relationship, Part 44 refers to product structure and configuration data, Part 45 refers to Materials, Part 46 refers to visual representation and drawings, Part 47 considers shape tolerances, Part 48 pertains to form features, and Part 50 to Process Planning.

The integrated application resources portion of ISO 10303 consists of: Part 101 referring to Drafting, Part 102 - Ship Structures, Part 104 - Finite Element Analysis, and Part 105 - Kinematics and Composites.

The Application Protocols portion of ISO 10303 consists of: Part

201 - Explicit drafting, Part 202 - Associative drafting, Part 203 - Configuration controlled design, Part 204 - Mechanical design using boundary representation, Part 205 - Mechanical design using surface representation, Part 206 - Mechanical design using wireframe representation, Part 207 - Sheet metal dies and blocks, Part 208 - Life cycle product change process, and Part 210 - Composites design, which includes electrical design applications.

EXPRESS. The EXPRESS language has been designed to define information models and is a component of the STEP Tools section of ISO 10303. The top construct in the EXPRESS language is a schema, which is made up of entities. One writes schemas in the following form:

```
SCHEMA nameOfYour Schema;
(* Entity Definitions *)
END_SCHEMA;
```

The entities are the constructs which define how the information is stored, exchanged, and structured. An entity has a name and a set of attributes, each of which, in turn, has a name and a data type. Entities are defined in the following manner:

```
ENTITY nameOfEntity;
AttributeName1: attributeDataType1;
AttributeName2: attributeDataType2;
.
.
.
AttributeNamen: attributeDataTypen;
END_ENTITY;
```

Each entity can have subtypes. A subtype is an entity that describes some subset of the entity. The usefulness of subtypes comes about because a subtype may have attributes which are not possessed by all instances of the supertype entity, but shared just by those in the subset.

The subtypes are declared as follows:

```
ENTITY nameOfEntity2
SUBTYPE OF (nameOfSuperEntity);
(*all the attribute definitions *)
END_ENTITY;
```

The existence of Subtypes implies the existence of Supertypes. An entity does not have to be declared Supertype of others, the subtype declaration suffices. On the other hand, it may be that one wants to make the supertype relationship explicit and avoid making an entity a possible supertype of some other entity. In that case, one should use the Supertype declaration. The EXPRESS keyword is SUPERTYPE.

```
ENTITY superEntity
SUPERTYPE OF (subentity1, subentity2,...,subentityn);
(* specifics of the superEntity attributes *)
END_ENTITY;
```

The primitive Data Types in EXPRESS are: REAL, INTEGER, BOOLAN, LOGICAL, STRING, and BINARY. The REAL data type can be given a precision that refers to the minimum number of resolved digits, for example:

```
length: REAL(15);
```

STRING represents a sequence of zero or more characters. Normally, the length of the STRING can vary. One may impose a maximum length by giving it a number within parentheses. If the specification is followed by the (reserved) word FIXED the length can not vary. For example:

```
name: STRING;
name2: STRING(10);
name3: STRING(15) FIXED;
```

The LOGICAL type can have three values: TRUE, FALSE or UNKNOWN.
The BOOLEAN type can have only two values: TRUE or FALSE. The BINARY type consists of a sequence of zero or one bits. As in the STRING case, a variable number of bits can be specified, a maximum number of bits indicated by the number in parentheses, or a fixed number of bits if the word FIXED follows in the specification.

In addition to the primitive data types, EXPRESS has four aggregate types: LIST, ARRAY, BAG, and SET. The first two are ordered arrays of elements, while the latter two are unordered. ARRAYs admit null elements, while the other three types don't. Duplicate values are allowed by all types except SET. ARRAYs have fixed sizes. Examples of the required syntax follow:

```
attributeName: ARRAY [1:5] OF REAL;
```

The above expression indicates an Array of up to 5 real numbers.

```
attributeName: ARRAY [1:5] OF OPTIONAL REAL;
```

The OPTIONAL keyword implies that some elements in the Array may not have a value (i.e., they could be NULL).

```
attributeName: ARRAY [1:5] OF UNIQUE INTEGER;
```

In this case, every element is different from another in the aggregate.

```
attributeName: BAG OF REAL;
```

This construct defines a Bag which can have zero or more elements in it.

```
attributeName: BAG [1:7] OF REAL;
```

The line above indicates that the BAG contains from 1 to 7 elements.

```
attributeName: BAG [1:?] OF REAL;
```

This definition declares that the BAG has at least one element.

```
attributeName: LIST OF REAL;
```

This is the way to define a list of zero or more elements.

```
attributeName: LIST[1:?] of REAL;
```

In this case we imply the LIST contains at least one element.

```
attributename: LIST[1:7] of UNIQUE REAL;
```

This construct indicates that we have a List of 1 to 7 distinct elements.

```
attributeName: SETOF REAL;
```

This declares a SET of zero or more elements.

```
attributeName: SET[1:7] OF REAL;
```

This construct implies a SET that has 1 to 7 elements.

```
attributeName: SET[1:?] OF UNIQUE REAL;
```

This refers to the case where the SET has 1 or more elements.

It is possible to use entities as data types. In addition, in EXPRESS one has the notion of a union of data types. This is indicated in the language through the use of the keyword SELECT. For example:

```
TYPE
SomeFigure = SELECT
(Square, Ellipse, Triangle)
END_TYPE;
```

```
ENTITY theFigure;
figure: SomeFigure;
origin: Point;
END_ENTITY;
```

As can be inferred from this example, the entity theFigure has an attribute named figure which can be any of three distinct entities: Square, Ellipse, or Triangle.

In addition to the TYPE declaration, it is possible to introduce enumerated data types, which are ordered sets of values which are represented by names. Each of these names must be unique within the enumerated set. As an example, consider:

```
TYPE color =
ENUMERATION OF (red,orange,yellow,green,blue,
violet,black,white,other,none);
END_TYPE;

ENTITY coloredFigure;
figure: SomeFigure;
origin: point;
lfill: color;
END_ENTITY;
```

In a given schema, it is possible to access entities defined in other Schemas, but one can not nest schema definitions. To have access to entities defined in other schemas use the words: REFERENCE FROM or USE FROM. The former construct allows access to the foreign schema entity definition, but continues to consider the definition as external, while the latter acts as if the foreign entity definition is now local, making it possible, among other things, to subtype the foreign entity. For example, let us assume someone has defined an entity in Schema Big, called hardEntityToDefine, and we want to use that definition within our Schema, called Ours.

The situation is as follows:

```
Schema Big;
.
(* Big Schema definitions *)
.
ENTITY hardEntityToDefine;
(* hardEntityToDefine definition *)
END_ENTITY;
(* more Big Schema Definitions *)
.
END_SCHEMA;


Schema Ours;
.
(* Ours Schema stuff *)
.
USE FROM Big (hardEntityToDefine AS hardMadeE-
asyAndLocal);

(* more Ours Schema, which incorporates use of hard-
MadeEasyAndLocal as it were just another locally defined
entity *)

END_SCHEMA;
```

If we wanted to incorporate all of Schema Big into Ours, we would just state:

```
USE FROM Big;
```

This appendix constitutes an introductory exposure to EXPRESS. There is much more to the language than what has been described here, especially in the areas of Rules, Functions, and Procedures. There is also a subset of the language, called EXPRESS-G, which defines graphical representations for entities and other constructs. The serious reader is encouraged to learn more about EXPRESS by reading and becoming familiar with the full language definition document [2.7]

2.6 References

[2.1] J. Rumble Jr. and J. Carpenter Jr., "Materials `STEP' into the Future," *Advanced Materials & Processes*, pp. 23-27, October, 1992.

[2.2] H. Moreno and S. Stark, "An MCM/Chip Concurrent Engineering Validation," MCC Technical Report HVE-042-93, March, 1993.

[2.3] "Reference manual for the ROSE++ data manager," STEP Tools, Inc. Rensselaer Technology Park, Troy, NY, 12180, 1992.

[2.4] H. W. J. Rittel, and M. M. Webber, "Dilemmas in a General Theory of Planning," *Policy Sciences 4*, pp. 155-169, 1973.

[2.5] W. Kunz and W. J. Rittel, "Issues As Elements of Information Systems," Working Paper No. 131, Institute of Urban and Regional Development, University of California at Berkeley, July, 1970.

[2.6] Documentation on STEP can be obtained from ISO, National Institute of Standards and Technology, Gaithersburg, MD. Refer to ISO TC184/SC4.

[2.7] "ISO, EXPRESS Language Reference Manual, External Representation of Product Definition Data," *ISO TC184/SC4/WG5 N14*, 1991.

Chapter 3

Tradeoff Analysis

3.1 Introduction

Tradeoff analysis is the process of comparing performance gains in one region of the design space with the associated performance losses in another. When simply enough systems are considered or the design constraints are strong enough, tradeoff analysis may lead to design optimization. However, in general, tradeoff analysis is performed at a "what if" level.

Figure 3.1 shows the interdependency amongst the various categories of metrics which must be considered during tradeoff analysis. It is obvious from Figure 3.1 that the problem can not be solved by simple superposition of prediction results; rather it is a problem of simultaneity. For example, adding wiring layers to an interconnect (substrate) might result in a smaller (area) module for an application, the smaller area module may however be more difficult to cool because the components are closer together requiring the addition of thermal vias in the interconnect, the thermal vias, in turn, block wiring tracks in the interconnect reducing the wiring capacity of the module sufficiently that the area advantage gained through the addition of extra layers is negated. Alternatively, suppose the additional thermal vias were not required in this example and the designer obtained the desired module shrinkage, but the module reliability is reduced as a result of the increase power dissipation density. The point of this example is that simple design changes sometimes do not produce the desired results, or they produce the intended result at the expense of reduced performance in another region of

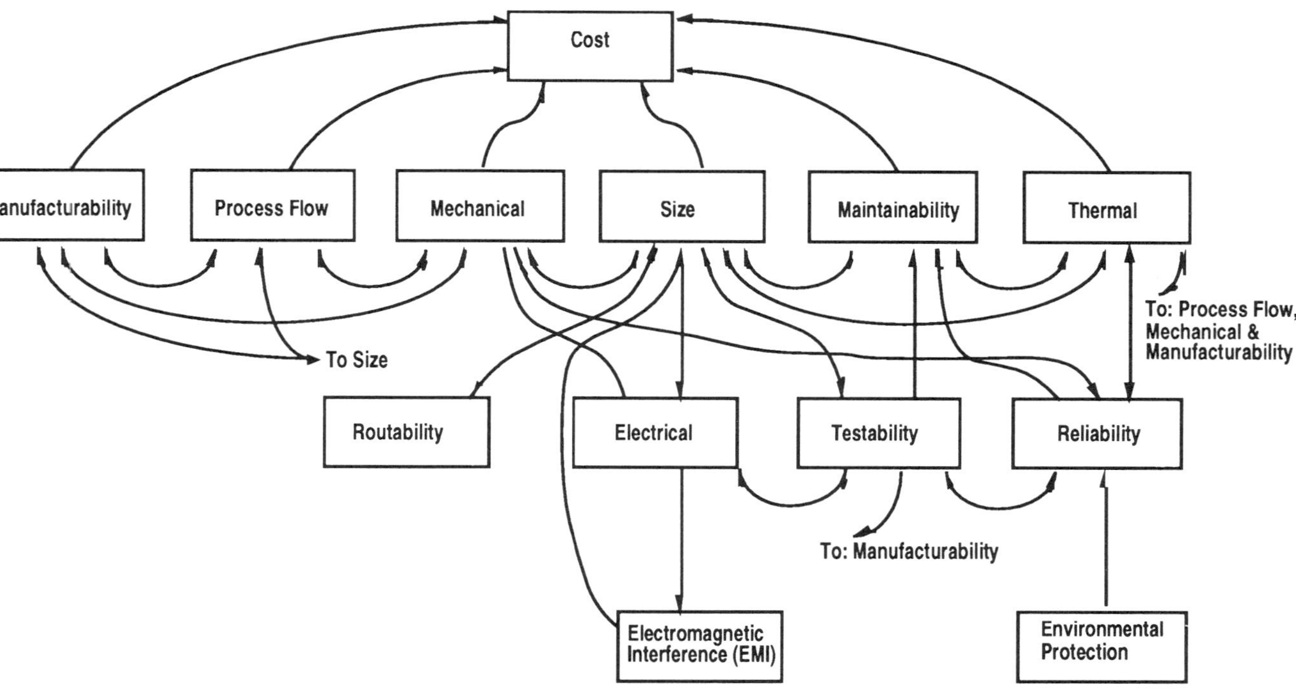

Figure 3.1. Estimation metrics necessary for the analysis of module-level packaging and interconnect problems. The links describe the interdependency of the estimation metrics.

the tradeoff space.

Figure 3.1 demonstrates a number of the significant problems encountered in performing tradeoff analysis for complex electronic systems:

1) The tradeoff space is extremely large. Assume, for example, that for a given module design there are four component set variations, four different candidate interconnect technologies, four different candidate bonding technologies for the bare components, and four different heat exchanger technologies to consider. If the final design must use one of each of these variations, there are a total of $4^4 = 256$ possible combinations to select from. Now consider the fact that all chips need not use the same bonding technology or the large number of other design and design rule tradeoffs which enter into the problem. The tradeoff space quickly becomes too large to search in a brute force fashion.

2) All the design views shown in Figure 3.1 are interdependent. It is dangerous to consider any one view alone.

3) It is unclear how to evaluate partial designs. One can not always infer a system's performance from the performance of a fragment of the system. Advantages gained for small portions of a system may be lost when the overall system is considered.

The tradeoff problem is further complicated by the existence of varying levels of designer supplied constraints and descriptions. While it is clear what information must be from chip design activities (i.e., physical descriptions of chips, power dissipations, noise margins, I/O count and distribution, driver and receiver models, etc.) the optimum abstraction level of their initial design descriptions varies considerably from design problem to design problem. Tradeoff analysis must be performed for mixtures of components consisting of well described die to packaged components which are "black boxes," as well as designs where the component set is fixed and designs where the contents and design details of the components may be varied to obtain better solutions.

3.2 Approaches to Tradeoff Analysis

Tradeoff analysis can be approached in several ways. Tradeoff analysis can be addressed in advisory, estimation-based predictive, or simulation-based predictive ways. Design advice explains to the designer how viable options compare and tracks compatibility amongst selected choices. Design advise is usually implemented using knowledge-base approaches. Predictive methods for tradeoff analysis involve the estimation or simulation of performance metrics. Besides obtaining qualitative design advise, the designer can benefit from quantitative feedback on the implications of design choices. Unfortunately detailed simulation and physical design activities are often too complex and time consuming to be a viable part of a broad tradeoff performance evaluation, so simplified estimation metrics are needed to help make detailed tradeoff practical.

The term “Design Advisor” has been used extensively in the recent literature to describe CAD tools which aid designers at some point in a complex design process, primarily tradeoff analysis, however no consistent definition of a design advisor exists. Design advisors can serve two purposes: 1) Observe the state of a design and assess its appropriateness against a defined set of constraints, an activity that could be used throughout the design process, and 2) Supplement specification and synthesis activities at the conceptual design level (before physical design activities). The same tool(s) can be used to fulfill both purposes if managed in a general enough sense.

One must take care not to confuse specific “design advisors” with their management functions (Figure 3.2). Design advisors may be defined as consisting of two components. The first component is design “analysis”. Design analysis considers the present state of the design and returns some quantitative measure of its performance (i.e., cost, size, speed, etc.). No interpretation of the results is performed (only the implications of design decisions are returned). The second component is design “evaluation”. Design evaluation interprets the results of a design analysis using a knowledge-base approach. The design analysis aspect may exist without design evaluation, if this is the case it is left up to the designer to interpret the results.

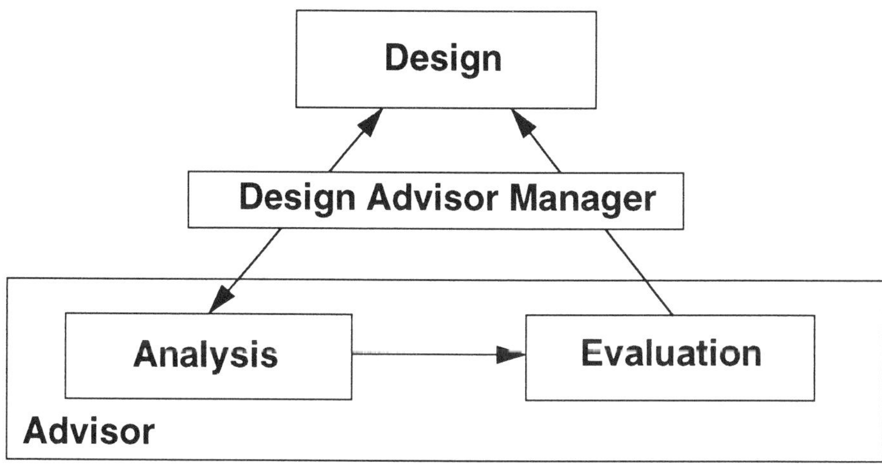

Figure 3.2. Design advisor components.

The design advisors never perform any actual design function, in the sense of altering the design. Changing the design is left up to the designer and synthesis tools. The design advisor "manager" manages the concurrent use of the multiple design advisors necessary to build a system-level view of a design.

3.2.1 Design Advice

Design advice presents the qualitative advantages and/or disadvantages that expert designers typically consider in accepting or rejecting a design option. Often, qualitative advice is sufficient for making a decision. There are however, cases in which the designer would like more detailed advice, i.e., how much better is one option than another? In this situation quantitative predictions are needed.

The concept of advice can be characterized by its type, basis, and ranking [3.1]. The type refers to the design issue which the advice is associated with. For example, a piece of advice associated with the area array flip chip bonding issue could be that flip chip bonding leads to low inductance bonds which will reduce switching noise but TCE (Thermal Coefficient of Expansion) mismatch between the die and the interconnect (substrate) becomes a problem for large I/O counts. Within a particular design issue, design advice is further characterized by discrimination factors which specify the basis by which options will be compared. In the context of this book, the discrimination factors take the form of performance metrics such as mean-time-to-failure, or maximum ground bounce. Finally, within a particular discrimination factor, design advice is further characterized according to its ranking. Ranking is an indication of the relative importance of the advice. Design advice may be ranked according to its quality (those performance metrics which are most accurately estimated could receive more weight), generality (those performance metrics which lead to the most general conclusions receive more weight), or constraint-driven (the user specifies the relative importance of one metric over another through the external entry of design weightings).

3.2.2 Prediction

Since actually executing a combination of tradeoff choices to determine if the resulting system meets performance targets and satisfies the necessary constraints is usually too cumbersome (and expensive) to be practical for most systems, predictive methods are often used to obtain early feedback on system performance. Predicted performance represents only an approximation, but is often sufficient to support initial high-level decisions.

There are a number of different types of predictive models [3.1]:

1) Heuristic Models. Heuristic models use general guidelines or empirical studies to model a system. A general guideline is a "rule-of-thumb." Example guidelines are shown in Table 3.1. The risk in basing prediction solely on general guidelines is that the guidelines can become outdated as technologies and materials change.

Heuristic models also make use of empirical study results. Quantitative empirical study results usually take the form of simple expressions obtained from fitting experimental or detailed simulation data for a specific situation (versus deriving an expression from basic principles). An example of empirical result is Rent's rule [3.3],

$$N_s = K_p N_g^{\beta} \tag{3.1}$$

Rent's rule relates the number of signal and control I/O to a circuit (or chip, or group of chips), N_S, to the number of subcircuits, (or gates, or chips), N_g, in the circuit. Rent's rule was originally observed by plotting I/O versus subcircuit count for real systems. From these plots the values of the constants K_p and β were originally determined[1].

Heat Transfer Modes	Design Characteristic
Natural Convection	• Applicable over moderate ambient temperatures • Nonlinear function of size, shape, orientation and temperature • Cools moderate to low component densities • Vented designs require louvers, screens or perforated surfaces • Improves with increasing surface area normally extended by fins, pins or decreased component density • Effectiveness decreases ~1.5% per 1000 ft increase in altitude
Forced Convection	• Requires supplied coolant or self-contained fan or pumps which could contribute to heat load • Applicable over moderate to high ambient temperatures • Cools moderate to high component densities • Requires the establishment of flow paths and baffles • Requires consideration of pressure loss and choking in selecting coolant prime mover • Requires consideration of acoustic noise
Conduction	• Cools moderate to high component densities • Requires a dedicated high pressure interface contact area • Requires use of high conductivity materials, short lengths and large cross sections
Radiation	• Requires consideration of surface finishes and coatings • Effectiveness independent of altitude increasing dramatically with greater temperature differences
Thermal Storage	• Requires materials with high thermal capacity • Requires large mass—which increases equipment weight

Table 3.1 - An example of heuristic guidelines. General guidelines for the use of various heat transfer modes [3.2]. (© 1985 Van Nostrand Reinhold Co.)

Heuristic models do not posses a "deep" understanding of the design process or the fundamental engineering principles and their validity, under circumstances other than the specific cases they were formulated from is not necessarily known. The advantage of heuristic models is that they require very little computational overhead and yield quick answers.

1. Since Rent's Rule was originally observed it has been derived from stochastic [3.4] and geometric [3.5] arguments.

2) Analytical Models. Analytical models use closed-form formulations derived from basic principles or probabilistic techniques to predict the performance of a system. Closed-form analytical expressions for predicting performance metrics can be derived from basic principles under a set of well understood assumptions. Because the assumptions are part of the model development, the applicability (and validity) of a model to a particular situation can be easily determined.

Analytical models based on probabilistic approaches involve capturing the characteristics of a design activity in a set of random variables with an associated set of probability density functions. An example of a probability based model is the wiring model developed by Sutherland and Oestreicher [3.6] discussed in the next section.

Analytical models tend to be more general than heuristic models, however, probabilistic models often lead to overly pessimistic results.

3) Simulation. Simulation is the detailed analysis of a critical portion of a design using approaches that usually consisting of solving a large number of simultaneous equations. Simulation generally results in the most accurate solutions, but often at the cost a large amounts of computation time and/or model setup time. As discussed in the Introduction (Chapter 1) it is often difficult to formulate a system level view large enough to aid in decision making without performing a very large number of simulations.

The designer is in no way bound to using a single predictive approach exclusively. Rather all approaches should be used in a hierarchical manner: The simplest, fastest, coarsest approaches are used initially on a large quantity of design variations followed by increasingly complex analysis methods applied to fewer designs, Figure 3.3.

3.3 The Metrology of Tradeoff Analysis

This section provides an in depth discussion of models and algorithms applicable to the conceptual design level. The performance metrics discussed here fit into the analytical model classification, however, in some cases analytical models are tied to detailed simulations to complete the analysis. Where possible the derivations leading to the estimation level formulations have been included and the region of applicability is discussed.

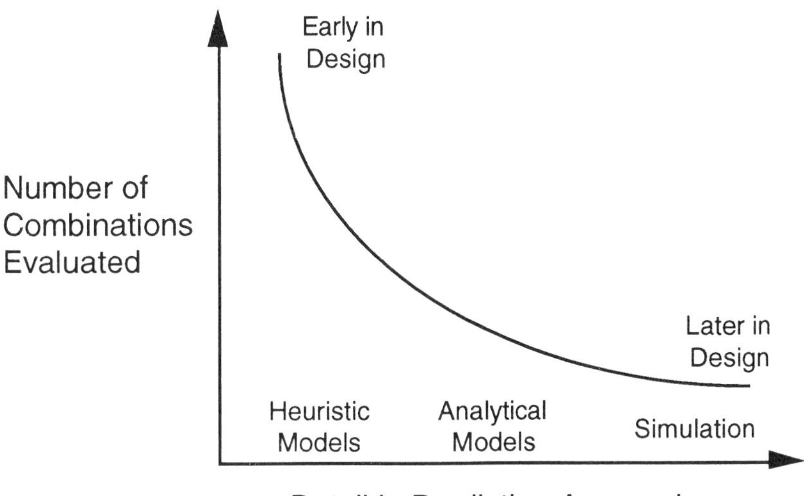

Figure 3.3. Hierarchy of predictive analysis methods.

3.3.1 Wiring (Routing) Analysis

One potential limitation of the size on complex electronic systems is the quantity of wiring resources available within an interconnect. System designers need to be able to predict the wiring utilization of electronic systems. A priori wirability (or routability) analysis can be accomplished using a hierarchy of techniques: 1) Heuristic-based estimations, 2) Statistically-based wiring simulators, and 3) Actual routing of test circuits using placement and routing tools. Wiring simulators produce fictitious wiring patterns which mimic actual wiring statistics. Actual routings require the development of netlists (whether actual or approximate) and the necessitate the entry of pin placement information; neither of these items are generally available in the early stages of conceptual design. In this treatment we are primarily interested in exploring heuristic-based wiring estimations.

Heuristic-based wiring predictions are probabilistic in nature, inasmuch as the precise knowledge of system architecture, partitioning, placement, and detailed interconnection is necessarily lacking at the early design stage when technology and design rule choices must be made. The goal of wiring prediction is to predict the required design rules and number of wiring layers to realize a design within user or system specified size constraints or to determine that wiring is not a size limiting constraint in a given problem.

Two approaches have been used to estimate wiring requirements. The first is based on the probability of wiring nets crossing selected cross-sections of boards. The second derives wiring limited component footprints by setting available wiring resource equal to wiring needs. Both approaches are described in this section.

The interconnect capacity (also called the connectivity) is a measure of the amount of wiring available in an interconnect. A common description of interconnect capacity is the available length of interconnect wiring per unit area of the board or module. The interconnect capacity (I_c) is described by

$$I_c = \frac{1}{W_p} N_w \tag{3.2a}$$

or

$$I_c = \frac{1+T_c}{V_p} N_w \tag{3.2b}$$

where W_p is the wiring pitch, N_w is the number of signal wiring or interconnect layers, V_p is the via or hole pitch, and T_c is the number of tracks or lines per channel. Equation (3.2a) may be applicable when vias are small or are not placed on a uniform grid and (3.2b) is the form which must be used when via or hole diameters are large and on a regular grid. The numerator of (3.2b) is T_c+1 instead of T_c to account for the track available to the channel grid points. The interconnect capacity given by (3.2) assumes that via pitches, tracks, and wire pitches are uniform over the whole area of a board or module. This is usually only approximately accurate. Wiring may be blocked by thermal vias, keepouts, and a host of other structures in real boards. The value of I_c should be modified to reflect these situations by subtracting the unusable wiring length.

Figure 3.4 shows the available interconnect capacities for the range of interconnect technologies in addition to those for bare die and wafer scale integration. Note that relatively large interconnect capacities can, and have been achieved using printed circuit board techniques. However, thin film multichip modules are capable of achieving equivalent or greater interconnect capacities with 2 wiring layers than printed circuit board or multilayer ceramic with many more wiring layers.

Some additional definitions are also helpful for the following discussion. A net is defined as a set of two or more pins (chip I/O or I/O leaving the module) which are electrically connected together. An interconnection is defined as a point-to-point connection. A net is made up of one or more interconnections. The fanout of a net is defined as one less than the number of pins a net connects together.

Section Crossing Estimations. The wiring limited size of a module or board, or alternatively the number of wiring layers, may be determined by estimating the number of connections which must pass through various board cross-sections. This approach was originally developed by Sutherland and Oestreicher [3.6].

Let P_1 and P_2 represent the fraction of signal pins on each side of an

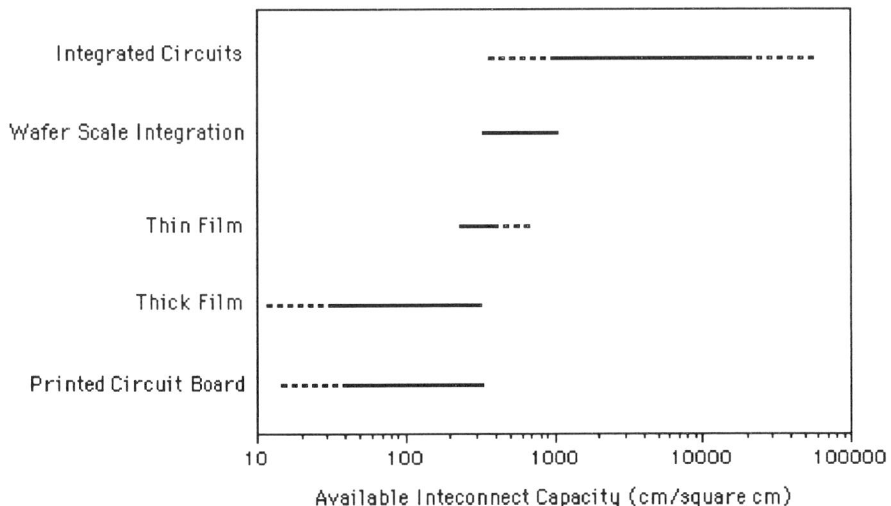

Figure 3.4. Typical available interconnect capacity for different technologies. Comparison of packaging technologies with interconnect capabilities on chip and using wafer scale integration.

arbitrarily drawn cross-section of a board or module so that $P_1+P_2=1$. A section is any line dividing the board into two parts (not necessarily in half). For randomly chosen nets, the probability that a net crosses the section at least once is given by

$$P_{cs} = 1 - P_1^n - P_2^n \tag{3.3}$$

where

n = the average number of pins per net (i.e. the fanout + 1)

P_1^n = the probability that the net starts on one side and connects only to pins on that side (never crosses the section).

The number of nets which cross the section is the probability of crossing multiplied by the total number of nets in the system (N_s/n), where N_s is the number of signal and control I/O,

$$N_{cross} = \frac{N_s}{n}(1 - P_1^n - P_2^n) \tag{3.4}$$

note that N_{cross} is a maximum when $P_1 = P_2 = \frac{1}{2}$. The minimum length of the section is then given by

$$L = \frac{N_w W_p}{F_i} N_{cross\,max} \qquad (3.5b)$$

or

$$L = \frac{N_w}{T_c + 1} \frac{V_p}{F_i} N_{cross\,max} \qquad (3.5b)$$

in this case F_i is the actual wiring efficiency (equations (3.5a) and (3.5b) correspond to (3.1a) and (3.1b)). In order to obtain the board area one should consider two perpendicular sections using the appropriate values of N_w,W_p, V_p, and T_c for each direction. The off-module connections can be accounted for by considering the module under analysis to be part of a larger board with the number of off-module connections equaling the number crossing between the module under analysis and the phantom larger board.

This method assumes random placement and would, for large arrays, yield excessively large estimates. Random placement means that a change in component placement can be expected to have no effect on the wiring density or average wire length on the board. This placement assumption is not applicable to circuits which have a high degree of regularity but can be used to provide an upper bound for them since using optimal component placement can only help not hinder. The theory predicts a minimum board size that should be easy to layout. Sutherland and Oestreicher have found that if moderately intelligent routing is combined with the random placement model, the expected number of wires crossing any center line of the board is 1/4 of the number of active circuit pins (signal pins) regardless of net size, and that the board area required per component increases as the number of components increases, i.e., big boards are less efficient than small boards.

Comparison of Requirements and Resources. The second class of wiring estimation approaches is based on setting the amount of resources available for wiring equal to the wiring requirements of the chips.

The amount of wiring resources available can be easily computed. The total wire length available is a function of the interconnect capacity given by (3.2), the fraction of that capacity which can be used for wiring (usually a maximum of 30-40%), and the size of the board or module which contains the wiring. The total wire length which can actually be used is given by

$$\text{total wiring length available for use} = \text{Area}F_iI_c \tag{3.6}$$

where F_i is the fraction of the available wiring that can be used (governed by the complexity of the routing problem and the quality of the routing approach), I_c is given by (3.2), and Area is the board or module area. If the chip set being considered is homogeneous (all the chips the same) then $\text{Area} = N_{chip}F_p^2$, where N_{chip} is the number of chips in the module and F_p^2 is the wiring limited footprint of one of the chips.

A simple wiring estimation which uses (3.6) was developed by Seraphim [3.7]. In Seraphim's approach the required length of wiring per chip site is derived in the following way. Assume a homogeneous set of components to be connected. The average connection length from a component to its nearest neighbors is F_p (the wiring limited chip footprint dimension) (see Figure 3.5). The Manhattan distance to the next nearest neighbors has an average length of $2F_p$. The average distance is taken as $1.5F_p$. The total wiring length per site is then given by $1.5F_p$ multiplied by the number of lines associated with the site. Assuming the average fanout is 1.5 and half the computed wiring is associated with the chip of interest (and half with all other sites) we obtain:

$$\text{required wire length per chip site} = 1.5F_p\frac{1.5N_s}{2} \tag{3.7}$$

In order to estimate the module size, the number of chip sites multiplied by (3.7) is set equal to (3.6) yielding,

$$L^2 = N_{chip} F_p^2 = \frac{N_{chip}}{F_i I_c} \tag{3.8}$$

In Seraphim's original derivation $F_i = 0.5$ was assumed. Equations (3.7) and (3.8) assume that all the sites are identical. This derivation is also limited by many other simplifying assumptions.

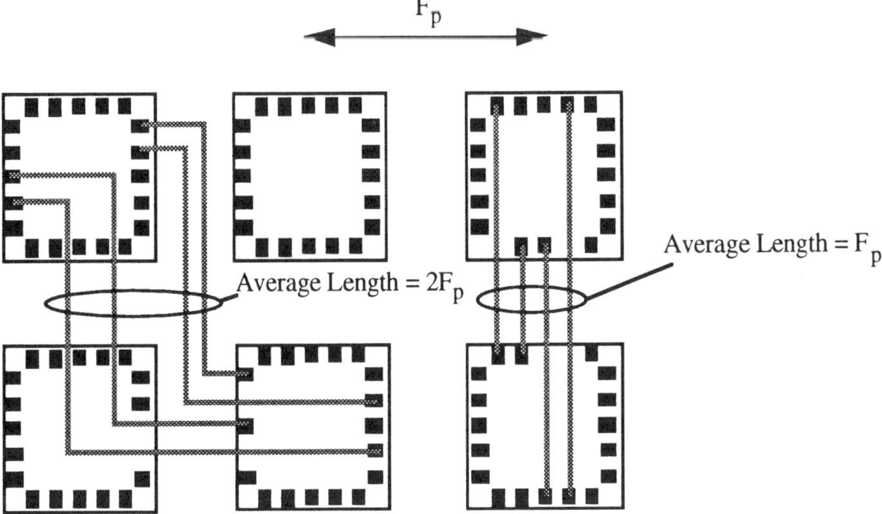

Figure 3.5. Definition of the wiring limited chip footprint dimension.

While Seraphim's model is simple and useful its assumptions about average interconnect length and net fanout limit its usefulness. A number of efforts have focused on the estimation of average interconnect lengths based on fits of real wiring data. These approaches generally derive interconnect lengths using Rent's Rule [3.3]. The best known work is that of Donath [3.8]. Donath derived a formula for an upper bound on expected average interconnection length based on partitioning results for linear and square arrays of gates. This formulation takes into account to a first order, the effect of placement on distance. The average wire length in units of gate pitch derived by Donath is,

$$R_m = \frac{2}{9}\left(7\frac{N_g^{p-0.5}-1}{4^{p-0.5}-1} - \frac{1-N_g^{p-1.5}}{1-4^{p-1.5}}\right)\frac{1-4^{p-1}}{1-N_g^{p-1}},\ p \neq 0.5 \qquad (3.9a)$$

$$R_m = \frac{2}{9}\left(7\log_4 N_g - \frac{1-N_g^{p-1.5}}{1-4^{p-1.5}}\right)\frac{1-4^{p-1}}{1-N_g^{p-1}},\ p = 0.5 \qquad (3.9b)$$

where N_g is the number of gates and p is the Rent's constant determined from experimental results for on-chip wiring length ($0 \leq p \leq 1$). The variable p in the above calculation can be interpreted as the degree of parallelism of the logic complex (large p = highly parallel), so placement distance is related to machine organization as well. This bound gives significantly lower interconnection length than a bound based on random placement.

Bakoglu [3.9] applies Donath's result at the chip level to the module level. By replacing the number of gates with the number of chips and the gate pitch with the chip wiring limited footprint dimension, F_p, the average interconnect length at the module level in units of F_p becomes,

$$R_m = \frac{2}{9}\left(7\frac{N_{chip}^{p-0.5}-1}{4^{p-0.5}-1} - \frac{1-N_{chip}^{p-1.5}}{1-4^{p-1.5}}\right)\frac{1-4^{p-1}}{1-N_{chip}^{p-1}},\ p \neq 0.5 \qquad (3.10a)$$

$$R_m = \frac{2}{9}\left(7\log_4 N_{chip} - \frac{1-N_{chip}^{p-1.5}}{1-4^{p-1.5}}\right)\frac{1-4^{p-1}}{1-N_{chip}^{p-1}},\ p = 0.5 \qquad (3.10b)$$

The value of the average interconnect length, L_{avg} is given by $L_{avg} = R_m F_p$.

Following Bakoglu's approach the interconnect capacity limited chip footprint is formulated by setting the total wiring length required to connect all the chips equal to the total wiring length available. The total wiring length required is

$$\text{total wiring length required} = \frac{f}{f+1} N_s L_{avg} \qquad (3.11)$$

where f is the average fanout of a chip's I/O, N_s is the total number of signal and control I/O in the system (all the chip I/O plus all the I/O going off the module), and L_{avg} is the average interconnection length in the module.

Setting (3.6) equal to (3.11) and solving for the chip footprint dimension gives,

$$F_p = \frac{f}{f+1}\frac{N_s R_m}{N_{chip} F_i I_c} \quad (3.12)$$

Similar to (3.8) the module area is given by $L^2 = N_{chip} F_p^2$.

Hannemann [3.10] has developed a wiring correlation similar to Bakoglu's. In Hannemann's formulation the module area is computed as,

$$L^2 = \left(c \frac{bN_s}{N_w \sqrt{N_{chip}}} \right)^2 \quad (3.13)$$

where

c = constant (suggested value = 3.9)

b = feature size parameter = $\frac{N_w}{I_c}$.

The correlation between Bakoglu's approach and Hannemann's can be obtained by setting L^2 in (3.13) equal to $N_{chip} F_p^2$ in (3.12) and solving for c. c is given by

$$c = \frac{f}{f+1}\frac{R_m}{F_i} \quad (3.14)$$

Figure 3.6 shows the correlation between the two models. It can be seen that the value of c = 3.9 yields a reasonable approximation to Bakoglu's model for modules with 10 to 30 chips and an average net fanout of 1.5 to 2.

An alternative form of this type of wiring estimation has been developed by Moresco [3.11]. Moresco's approach is based on a requirements = resources argument similar to Bakoglu's, but is not Rent's rule based, rather it is an extended neighbor counting based method similar in concept to Seraphim. In Moresco's approach the wiring resources are computed as in (3.6). The wiring requirement is computed by assuming

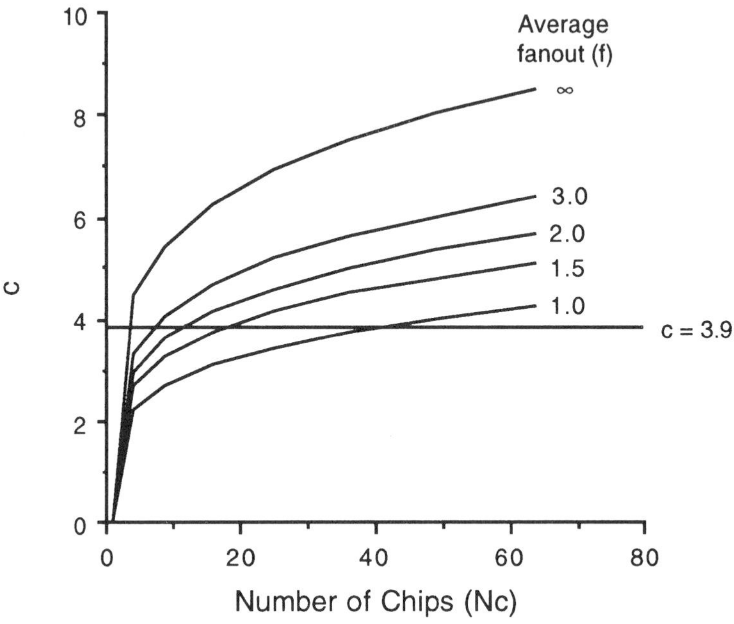

Figure 3.6. The correlation between Hannemann's and Bakoglu's wiring estimations.

that a fraction of the nets (A) are nearest neighbor routed and the rest of the nets (1 - A) are globally routed. These two routing extremes are explained in Figure 3.7. In this approach the total required wire length is given by,

$$\text{total wiring length required} =$$

$$A\,\frac{N_{chip}N_{sc}F_p}{2} + (1-A)\,(N_{chip}-1)\,N_cF_p + \frac{\sqrt{N_{chip}}\,N_{se}F_p}{v} \tag{3.15}$$

where,

N_{sc} = the number of signal and control I/O per chip

N_{se} = the number of signal and control I/O leaving the module

v = 2 for edge connector and 4 for an area array connector on the bottom of the module.

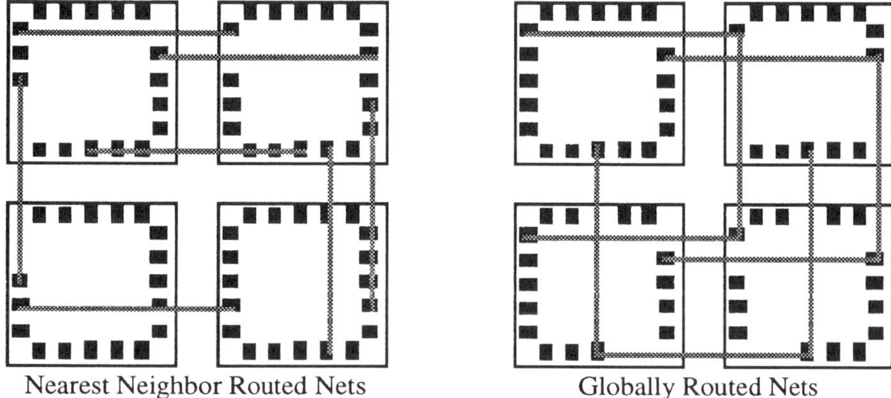

Figure 3.7. Nearest neighbor and global routing extremes.

The first term in (3.15) represents the fraction of nets which are nearest neighbor routed. This term can be derived from (3.11) by setting $R_m = 1$, ($L_{avg} = F_p$) and the average net fanout, f = 1. The second term in (3.15) represents the fraction of nets which are globally routed. This term can be derived from (3.11) by setting $R_m = 1$, ($L_{avg} = F_p$) and the average net fanout, $f = (N_{chip} - 1)$. R_m is 1 in this case (not $N_{chip} - 1$) because R_m is the average interconnect length not the average net length. The last term in (3.11) represents the wiring length necessary to attach all the chips to the off-module I/O. To derive this term from (3.11) the average net fanout is assumed to be one (f=1); for an edge connector $R_m = \sqrt{N_{chip}}$ $(L_{avg} = F_p\sqrt{N_{chip}})$ and for an area array connectorization scheme $R_m = \frac{\sqrt{N_{chip}}}{2}$, $\left(L_{avg} = F_p\frac{\sqrt{N_{chip}}}{2}\right)$.

All of the approaches discussed in this section, in their simplest form, assume that the component set is homogeneous (i.e., all the components are the same). This is usually not a good approximation to real world problems. It is important to note that this restriction appears at two critical points, 1) in the derivation of the wiring capacity limited footprint, and 2) in how the overall module size is determined (treated in the next section. One method of fixing this problem is to recompute an effective number of chips and a corresponding average interconnect length for each component. The simplest way to compute a new effective number of chips is from the ratio of the number of signal and control connections in the whole module and the number of signal and control connections that the ith component has,

$$N_{chip_i} = \frac{\text{Number of signal \& control connections in the whole module}}{\text{Number of signal \& control connections in chip i}} \tag{3.16}$$

If this method is used it is no longer sufficient to simply multiply F_p^2 by the number of chips to get the module area, instead a summation must be used,

$$L^2 = \sum_{i=1}^{N_{chip}} F_{p_i}^2 \tag{3.17}$$

Alternatively, Schmidt [3.12] has generalized Donath's treatment to allow for nonuniform subpackage size and terminal count.

Statistically-Based Wiring Simulators. So far this section has review heuristic-based wirability estimations. These approaches can only estimate the amount of wiring needed for a particular problem. They however do not address whether the problem is in fact wirable. In order to predict the wiring difficulty of a particular problem either a wiring simulator must be used or the problem must actually be routed.

Wiring simulators require more detailed information about the wiring problem then the simple heuristic-based estimations. Wiring simulators target emulating a more detailed router tool. The inputs to these more detailed models include: a rough netlist describing the major global connections (i.e., data paths and buses), attributes of the major connections (type - clock, dc, etc., performance, driver type, driver characteristics), wiring strategy (assignments of particular nets to specific layers of a board), and wiring topology (distributed, near-end cluster, far-end cluster, etc.). Wiring simulators can produce fictitious wiring patterns that mimic actual wiring statistics (average wire length, wire length distribution, via usage). Wiring simulators can also perform congestion analysis and chip placement evaluations. An advanced wiring simulation package suitable for conceptual design analysis is contained within the PEPPER tool developed at IBM [3.13].

3.3.2 Sizing Analysis

Almost all measures of system performance and manufacturability will in some way be affected by the size of the module. The size of the module can determine the length of critical nets, which may be an important factor in determining system cycle times and other electrical performance issues. How closely the chips are packed will determine the power dissipation density of the module which may determine how effectively or economically the module can be cooled. The size of the module will also contribute to its reliability and yield which affect the cost. The size of the module not only depends on the size and number of components that must be placed on or in it, but also on the requirements placed upon the connection of the components.

Module size is determined by a collection of effects which include wiring capacity, routability, bond pad pitch, escape routing, and placement, in addition to the actual physical sizes of the components being interconnect. The last section focused on only the wiring capacity limitation; in this section this limitation is considered along with the additional factors listed above in order to assess actual module sizes.

The discussion in the preceding section considered only the availability of enough wiring to route the system but did not treat the accessibility of that wiring from the die, or the many other factors which may contribute to determining the actual size of a system.

Bond Pad Pitch and Via Density. The footprint of a component can be limited by the allowable minimum pitch of bond pads on the interconnect and/or the density of vias available to take signals to various wiring layers. The bond pad pitch limited footprint can be computed for peripheral and area array components using,

$$\text{Area Array: } F^2_{p(\text{bond pad density})} = N_c B_p^2 \tag{3.18a}$$

$$\text{Peripheral (single row): } F^2_{p(\text{bond pad density})} = \left(\left(\frac{N_c}{4} + 2\right) B_p\right)^2 \tag{3.18b}$$

$$\text{Peripheral (double row): } F^2_{p(\text{bond pad density})} = \left(\left(\frac{N_c}{8} + \frac{5}{2}\right) B_p\right)^2 \tag{3.18c}$$

where N_c is the total number of I/O possessed by the component and B_p is the minimum bond pad pitch. Area array refers to bond locations spread uniformly across the face of the die and peripheral denotes bonding locations around the die edge in a single or double row. Equations (3.18b) and (3.18c) assume four sided bonding and square bond pads. Figure 3.8 shows the assumptions used to formulate (3.18b) and (3.18c). Note that (3.18b) is only exact when N_c is a multiple of 4 and (3.18c) is only exact when N_c is a multiple of 8 plus 4. More detailed versions of (3.18), which account for rectangular bond pads and different pad placement schemes can be derived.

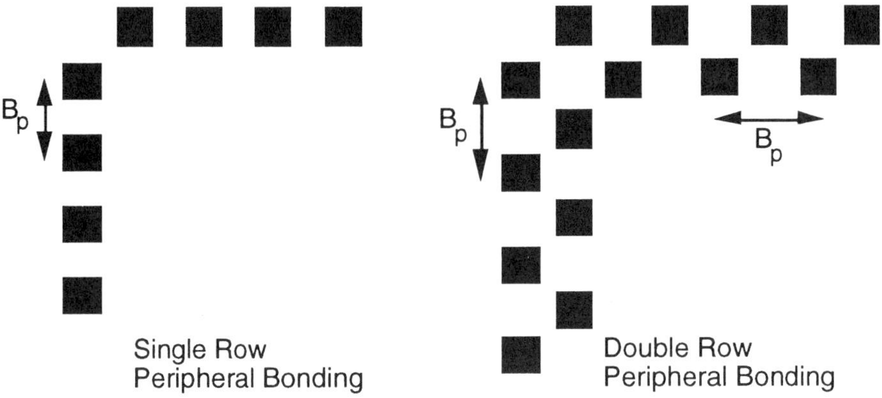

Figure 3.8. Single and double row peripheral bonding assumptions.

Figure 3.9 shows how the bond pad limited footprint varies as a function of the number of I/O on the die. Each line on Figure 3.9 represents the minimum footprint size possible with the particular bonding approach and design rules. As the number of I/O increases the minimum area required by the peripheral bonding approach, as given by (3.18), increases more quickly than that required by the area array approach. The dashed line in Figure 3.9 represents minimum physical die sizes computed from Rent's rule[2]. The footprints of chips that follow Rent's rule

2. Equation (3.1) for CMOS microprocessors with $K_p = 0.82$ and $\beta = 0.45$, and (3.10) with p = 0.4. Peripheral bonding is assumed and a 4 to 1 signal to ground ratio was used.

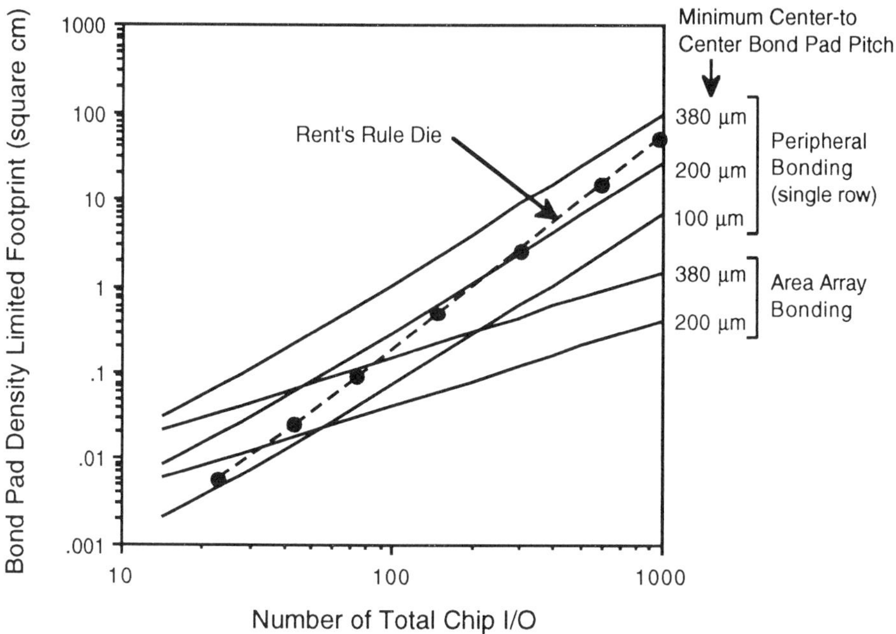

Figure 3.9. Bond pad density limited footprints for various peripheral and area array pitches. All the footprints were evaluated using (3.18). The dashed line indicates die sizes for Rent's rule limited microprocessors assuming 1 μm design rules.

will not be limited by the bond pad density for cases below the dashed line and may be limited for cases above.

Another component footprint measure closely related to the number of I/O is the I/O density. I/O density is the ratio of the number of I/O to the area of the footprint which is allowed. Figure 3.10 shows the I/O density for a range of single chip packages and bare die bonding approaches. For all approaches except area array, the I/O density decreases as the number of I/O increases.

An important factor in determining packaging density is interconnect accessibility. Interconnect accessibility refers to the ability to connect to the internal wiring from the external bonding surfaces on which components are mounted. Interconnect accessibility is limited by the density of vertical connections (vias or holes) between the surface on which bare die or single chip packages are bonded and the interconnect

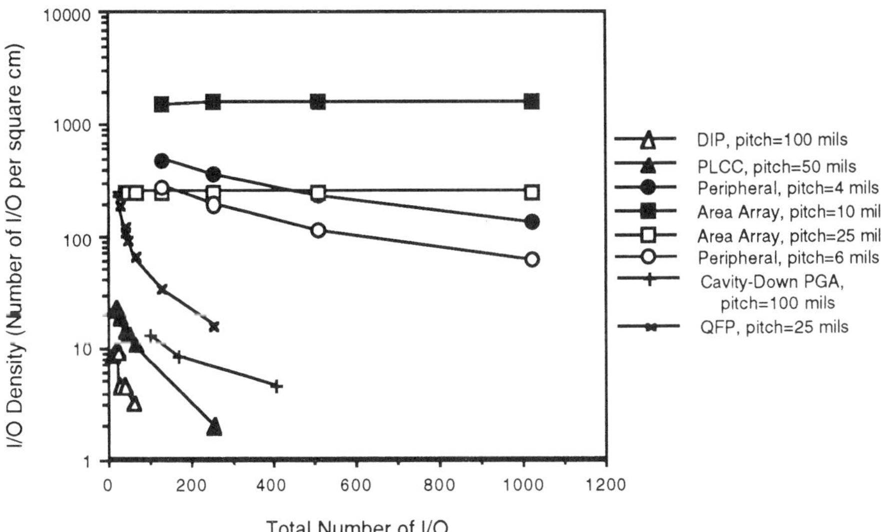

Figure 3.10. I/O density for various single chip packages and bonding methods. The I/O density is the ratio of the number of I/O to the package size (plus the area required by the bonds, if any). The peripheral (single row) and area array package areas were computed from (3.18). A bond length of 13 mm (50 mils) was assumed in all peripheral cases. In all cases sufficient interconnect capacity and via accessibility was assumed to exist.

layers, and the size and distribution of bonding pads. Thermal vias which are usually accompanied by heat spreaders under chips on the bonding layer also impact interconnect accessibility. The presence of heat spreaders and the use of vias or through holes for thermal purposes can reduce or preclude fanning traces from bond pads under chips in order to access electrical vias. If fanning traces outward from chips is necessary, then the physical footprint of the component may increase.

The via density limited footprint can be computed for components using,

$$F^2_{p(\text{via density})} = \frac{N_c}{F_v \rho_{via}} + \text{unusable area} \tag{3.19}$$

where N_c is the total number of I/O possessed by the component, ρ_{via} is the via or through-hole density (number of vias per unit area), and F_V is the efficiency of via use. F_V is used to account for vias which are present but not reachable. The "unusable area" appearing in (3.19) represents module surface area which is not available for via access due to heat spreaders or other structures.

Escape Routing. Escape routing refers to the translation of the I/O pads on the die or single chip package to lines which can escape from the area immediately surrounding the die to find vias, holes or top layer routing. Escape routing is a critical concern when the I/O on the die or single chip package are in an area array format (i.e., flip chip bonded bare die or ball grid array packages like OMPACs [3.14]). The footprint of peripherally (single row) bonded components may be limited by the number of vias available but will not be limited by escape concerns. Escape routing implies that for a given interconnect/bonding technology there is an upper limit to the number of I/O per unit area (I/O density) allowed.

The information necessary from an escape routing analysis is: can the die be escape routed and if so how many layers are necessary? The answers to these questions are technology and design rule dependent.

The most difficult escape case to consider is that of a uniform area array like the one shown in Figure 3.11 (i.e., an area array flip chip die or a ball grid array package).

The number of layers necessary for escape routing can be found using the following algorithm:

1) Assuming that there are no power or ground I/O in the outermost row the number of I/O which can be escape routed on the top layer of the interconnect is given by

$$n_{top} = 4\,(p_{top} + 1)\,(\sqrt{N_c} - 1) \qquad (3.20)$$

where p_{top} is the number of lines (tracks) allowed between pads on the top layer and N_c is the total number of I/O to be accommodated. Assuming that power and ground I/O are uniformly distributed within N_c and that they do not need to be escape routed, the total number of I/O accommodated by the top layer alone is,

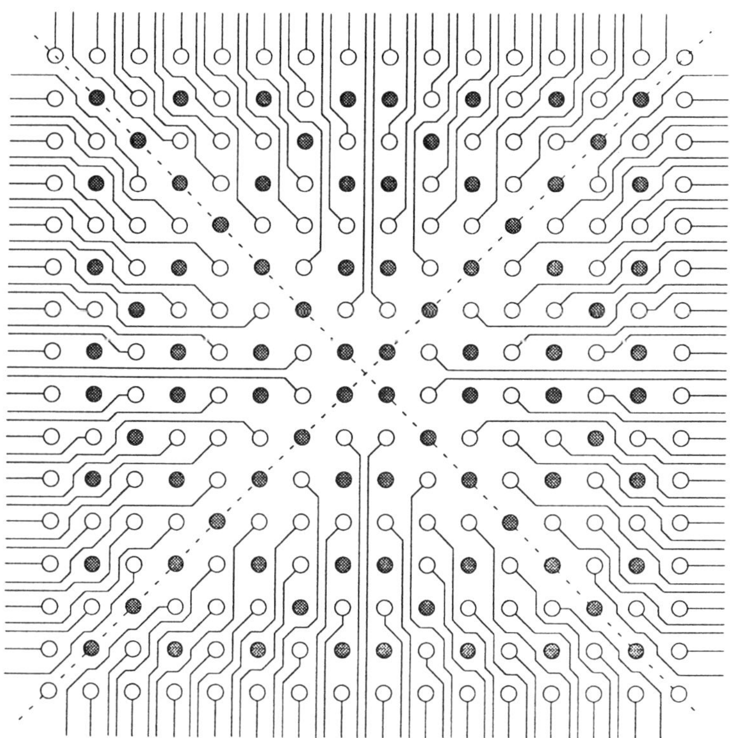

Figure 3.11. Possible escape routing from a uniform 16 x 16 area array of bond pads with two lines allowed between each pad. The shaded points are power and ground I/O. (Courtesy of Bill Weigler, MCC)

$$\text{I/O accommodated by the top layer alone} = n_{top} + \frac{2N_c}{2 + S{:}G} - \frac{8(\sqrt{N_c} - 1)}{2 + S{:}G} \qquad (3.21)$$

where S:G is the signal to ground ratio and an equal number of power and ground I/Os are assumed. The second term in (3.21) is the total number of power and ground I/O. The third term in (3.21) is the number of power and ground I/O in the outermost row (subtracted because the n_{top} calculation assumes that there are no power or ground I/O in the outermost row, if there are no power or ground

I/O in the outermost row the third term is not necessary). If the number of I/O accommodated on the top layer is greater than or equal to N_C then the component can be completely escape routed on the top layer. If the number of I/O accommodated on the top layer is less than N_C then the number of I/O remaining to be escape routed on other interconnect layers is given by

$$n_{remaining} = N_c - n_{top} - \frac{2n_{top}}{S:G} + \frac{8(\sqrt{N_c} - 1)}{2 + S:G}\left(1 + \frac{2}{S:G}\right) \tag{3.22}$$

The last term in (3.22) is not necessary if no power and ground I/O are in the outermost row.

2) If additional layers are required. The number of I/O which can be escape routed on the ith wiring layer of the interconnect is given by

$$n_i = 4(p_i + 1)\left(\sqrt{n_{remaining_{(i-1)}}} - 1\right) \tag{3.23}$$

where p_i is the number of lines (tracks) allowed between pads on the ith wiring layer and $n_{remaining(i-1)}$ is the remaining number of I/O after using i-1 layers. The number of I/O remaining to be escape routed on other interconnect layers is given by

$$n_{remaining_i} = n_{remaining_{(i-1)}} - n_i - \frac{2n_i}{S:G}$$

$$+ \frac{8\left(\sqrt{n_{remaining_{(i-1)}}} - 1\right)}{2 + S:G}\left(1 + \frac{2}{S:G}\right) \tag{3.24}$$

Layers should be added until $n_{remaining_i} \leq 0$.

The theoretical maximum array size for single layer escape routing can be found by setting (3.21) equal to the total number of I/O which must be accommodated,

$$N_c = 4(p_{top}+1)(\sqrt{N_c}-1) + \frac{2N_c}{2+S{:}G} - \frac{8(\sqrt{N_c}-1)}{2+S{:}G} \tag{3.25}$$

solving for $\sqrt{N_c}$ we obtain,

$$\sqrt{N_c} = \frac{b}{2}(-1-\sqrt{1+4/b}) \tag{3.26}$$

where $b = -\frac{4(p_{top}+1)(2+S{:}G)}{S{:}G}$ if power and grounds are not allowed in the outermost row of bond pads (such as in Figure 3.11), or $b = -\frac{4(p_{top}+1)(2+S{:}G)-8}{S{:}G}$ for the general case. Equation (3.26) is plotted in Figure 3.12 for the case where no powers or grounds are allowed in the outermost bonding pad row. In general the largest array sizes are possible with the smallest signal to ground ratio and largest number of lines between pads.

Other Footprint Limitations. Obviously the footprint of a component could also be limited by the physical size of the bare component or the physical size of the component in a single chip package or carrier. The physical footprints can be computed from,

$$F^2_{p(\text{die or single chip package})} = (L_{chip}+2L_{bond}+s)(W_{chip}+2L_{bond}+s) \tag{3.27}$$

where L_{chip} and W_{chip} are the dimensions of the die or single chip package, L_{bond} is the maximum bond length (zero if flip chip bonding is used), and s is the minimum spacing between the bonds or leads of this component and the bonds or leads on adjacent components.

A thermal limited footprint based on power dissipation density can also be used. This footprint takes the form,

$$F^2_{p(\text{power dissipation density})} = \frac{P_c}{\rho_p} \tag{3.28}$$

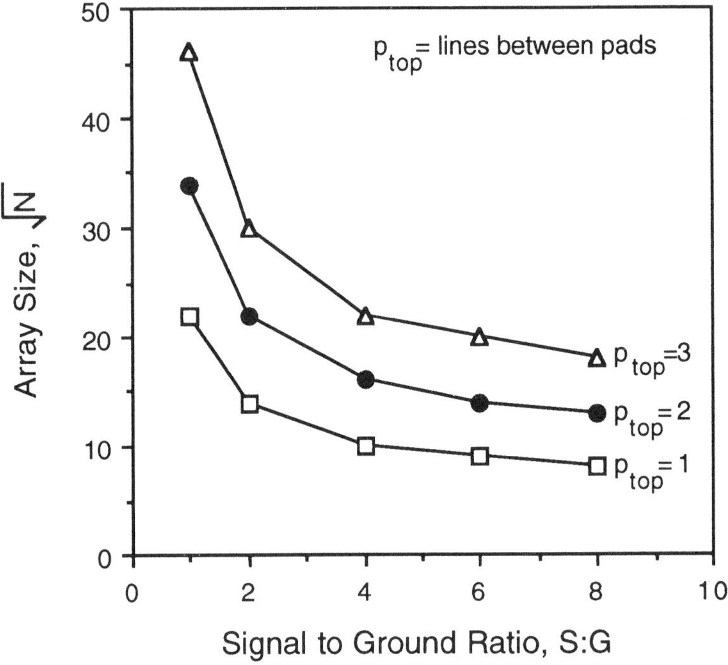

Figure 3.12. The theoretical maximum array size as a function of signal to ground ratio and the number of line allowed between pads for single layer escape routing (no power or ground I/Os in the outermost row).

where P_c is the power dissipation of the die and ρ_p is the allowed power dissipation density. ρ_p can be set by constraining the type of cooling strategy for the module. The relationship of footprint size to cooling requirements is treated in more detail in [3.15].

Additionally, a footprint limitation based on switching noise could be developed.

Module Size. The concepts developed so far in this section can be used to estimate the amount of module area a component requires (called the component's footprint). The footprint of a particular component is determined by the largest size limitation imposed by the factors,

- $F^2_{p(\text{interconnect capacity})}$
- $F^2_{p(\text{via density})}$
- $F^2_{p(\text{bond pad density})}$
- $F^2_{p(\text{single chip package})}$
- $F^2_{p(\text{die})}$
- $F^2_{p(\text{power dissipation density})}$

If all the components in the module are identical and attached to only one side of the interconnect, then (ignoring placement constraints) the minimum module area is the number of components multiplied by the maximum of the foregoing terms. For modules which contain chips of many different sizes and I/O counts, the formulation of the module area has to take into account that a chip whose footprint is limited by $F_{p(\text{die})}$ or $F_{p(\text{single chip package})}$ can be placed within the footprint of an adjacent chip who's footprint is limited by $F_{p(\text{interconnect capacity})}$. The possible placement of components on both sides of the module must also be considered. The general case can be approximately handled by accumulating the interconnect capacity limited footprints (and in the case of through-holes the via density footprints) of all the chips, separately, from the physical and bond pad density footprints of components. Note that components which are through-hole mounted require a physical footprint on both sides of the module or board. The physical and bond pad density footprints must also be accumulated separately on each side of the module if components are mounted on both sides. The largest size obtained from this process is called the minimum theoretical module area. The actual module area is the larger of the minimum theoretical area and the area obtained from a placement of the components.

In addition to the effects discussed above there is an additional module or board size constraint, namely, the connection to the next level of packaging. The module size limitations are given by (3.18) with N_c replaced by the number of off-module I/O and B_p replaced by the minimum bond pad pitch associated with the connector or connection scheme.

Fanout and Single Chip Packages. Suitably fanned-out devices will allow modules and boards with relatively modest interconnect capacities to perform complex wiring tasks; i.e., the sole future use of traditional leaded single chip packages would virtually negate all need for increasing the interconnect capacity of modules connecting chips in single chip packages.

In order to identify the interconnect capacity required in various situations, consider the following example, [3.16]. Ignoring the system/connector technology drivers and via accessibility constraints, consider just the interconnect capacity required by the various applications. Figure 3.13 shows the on-module I/O density as a function of the intercon-

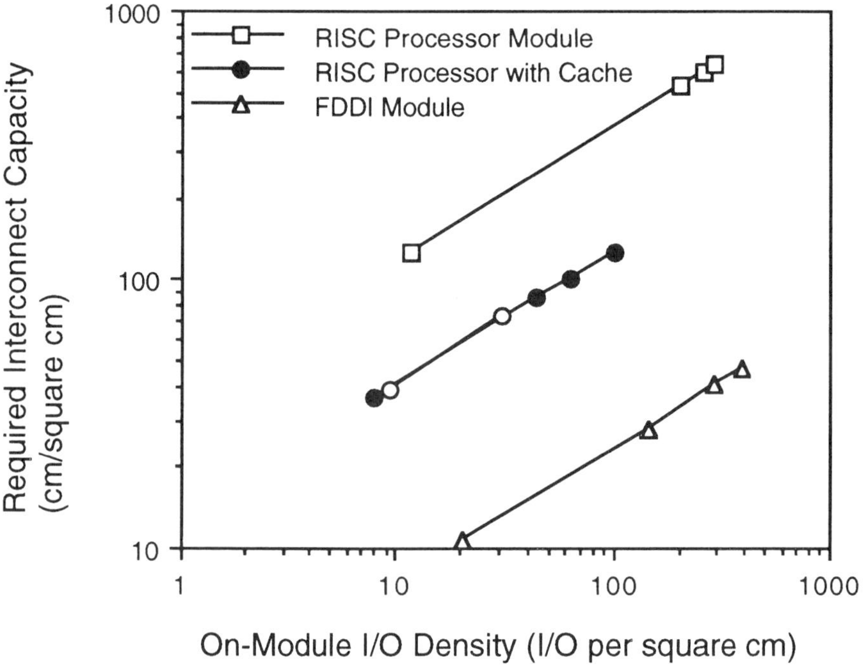

Figure 3.13. Required interconnect capacities for the three applications described in the text. The four data points included for each applications from left to right are: 1) single chip package solutions (all die in single chip packages), 2) bare die assembled using TAB or wirebonding, 3) bare die assembled using area array flip chip, and 4) chip-first (overlay) methodologies. The two open circle points on the line representing the RISC processor with cache memory denote surface mounted SRAMs with all other chips assembled as bare die using TAB or wirebonding (from left to right: SRAMs in SOJ packages and TSOPs)

nect capacity required such that the size of the module is not wiring limited. The interconnect capacity was determined by solving (3.12) for I_c under the assumption that enough wiring exists so that the size of the module is not interconnect capacity limited. In all cases a maximum of 40% of the available interconnect wiring was assumed to be actually usable due to non-optimal routing, imperfect floorplanning, and congestion. Results for three different modules are shown: a 6 chip RISC processor module containing four CPU chips and two FPUs, a 24 chip RISC processor module containing two VLSI chips in addition to instruction and data caches, and a third application representing a few--chip package application consisting of a portion of a fiber distributed data interface (FDDI) chip set with one physical layer controller (PLC) and a physical data transmitter (PDT) and receiver (PDR). More detailed descriptions of the components appear in [3.16].

Figure 3.13 shows that suitably fanned-out devices allow relatively modest interconnect capacity modules to perform complex wiring tasks, for example, sole future use of single chip packaged die would virtually negate all need for increasing the wiring capacity of interconnects beyond ~120 cm/cm^2 (300 inches/inch^2) even for very high density applications. In fact for realistic cases like the RISC processor with cache memory which contains a sizable number of passive components, the necessary wiring capacity when using single chip packages rarely exceeds 40 cm/cm^2 (100 inches/inch^2). Conversely, Figure 3.13 shows that adoption of advanced package approaches (bare die mounting or very small outline single chip packages) requires a commensurate increase in required interconnect capacity.

3.3.3 Thermal Analysis

Supporting high heat fluxes while maintaining relatively low component temperatures is one of the major challenges facing today's electronic system designers. Acceptable operational temperature levels are dictated by performance constraints, reliability demands, and the requirement that complex systems operate over a wide variety of environmental conditions.

Power is removed from an electronic system by heat transfer. In order for an electronic system to operate correctly the temperature of transistor junctions must be maintained below specified critical levels by conducting the heat dissipated by the device out of the system. Heat must be conducted from small hot areas to large cooler areas where the heat can be convected away either by a gas or liquid.

Electronic systems are becoming smaller through the use of advanced packaging techniques and multichip modules (MCMs), unfortunately the power dissipation of these subsystems is not decreasing (and is sometimes increasing). Thermal designers are therefore faced with the management of power dissipation densities which are increasing quickly. The continuation of this trend requires that thermal considerations begin at the conceptual design level and that cooling strategies be considered at the earliest possible stage of the design.

All microelectronic packages and structures can be characterized by their thermal resistance. Thermal resistance is defined as

$$R_t = \frac{\Delta T}{Q} \tag{3.29}$$

where ΔT is the temperature drop and Q is the heat flow. A high thermal resistance is associated with poor thermal conductors that undergo significant heating when energy is conducted through them. Low thermal resistances indicate good thermal conductors which undergo very little internal heating when energy is conducted through them. Ideally electrical packaging would be constructed from materials with zero thermal resistance so that 100% of the heat dissipated by the components could be conducted out of the system with no system heating.

The total thermal resistance in the heat flow path from a component to the outside of a system can be broken down into three elements [3.17]. 1) The internal thermal resistance (R_{int}) which exists between the junction where the heat is generated and the outside surface of the component case (through the top of a single chip package or through an interconnect). 2) The external thermal resistance (R_{ext}) which exists between the surface of the component case and some reference point. 3) The system level thermal resistance, the final temperature rise encountered when transferring heat from the coolant (gas or liquid) to some ultimate heat sink.

The junction temperature of a particular device is the most critical thermal parameter. The reliability and performance of a device are dependent on its junction temperature. The junction temperature is related to the thermal resistances in the thermal path by

$$T_j = P_c R_{int} + P_m R_{ext} + \Delta T_{coolant} + T_{coolant\ in} \qquad (3.30)$$

where,

P_c = the chip power dissipation
P_m = the module power dissipation
R_{int} = internal thermal resistance associated with the chip
R_{ext} = external thermal resistance associated with the module
$\Delta T_{coolant}$ = temperature rise of the coolant from inlet of system to module under consideration
$T_{coolant\ in}$ = inlet coolant temperature.

Thermal Resistance Estimation. The internal thermal performance of various assembly/interconnect variations is assessed by computing the thermal resistance through a selected path or paths. The discussion that follows focuses on a path "through the substrate" under each chip in a module. Note that the "substrate" is not necessarily located on the same side of the components as the interconnect's wiring in all cases (the chip-first thin-film overlay approach is one exception). The total internal thermal resistance associated with a component is the sum of the internal thermal resistances to ambient through all applicable paths (i.e., series and/or parallel paths through the module or board under the die and to the environment above the die). Often one of the two paths is ignored due to design constraints. The basic relation for thermal resistance is formed using (3.29) and the rate equation for conduction (Fourier's Law),

$$Q = -k_t A \frac{dT}{dx} \tag{3.31}$$

where,

Q = heat flow normal to area A
A = area
k_t = thermal conductivity
dT/dx = temperature gradient.

If the heat flow is steady, in one dimension, and a uniform bulk material is assumed then using (3.31), the thermal resistance in (3.29) becomes

$$R_t = \frac{t}{k_t A} \tag{3.32}$$

where t is the material's thickness, k_t is the material's thermal conductivity, and A is the cross-sectional area of the region through which heat flows. The total thermal resistance is computed by adding appropriate combinations of the thermal resistance through all materials in the selected thermal path from the junction dissipating the power to the "case," i.e., the die, die attach, bonds, heat spreader, dielectric layers, thermal vias, and substrate in series and parallel combinations as shown in Figure 3.14. Terms which account for spreading outside the area of the die and heat spreader, and constriction into thermal vias must also included.

An example of "bulk" thermal resistance is the thermal path through the material used to glue a bare die to an interconnect or into a single chip package. For a face-up bonded die (TAB and wirebond) the die attach thermal resistance is computed using,

$$R_{\text{die attach}} = \frac{t_{\text{die attach}}}{k_{\text{die attach}}\, \text{area}_{\text{die}}} \tag{3.33}$$

where,

$t_{\text{die attach}}$ = die attach thickness
$k_{\text{die attach}}$ = thermal conductivity of the die attach material
area_{die} = cross-sectional area of the die.

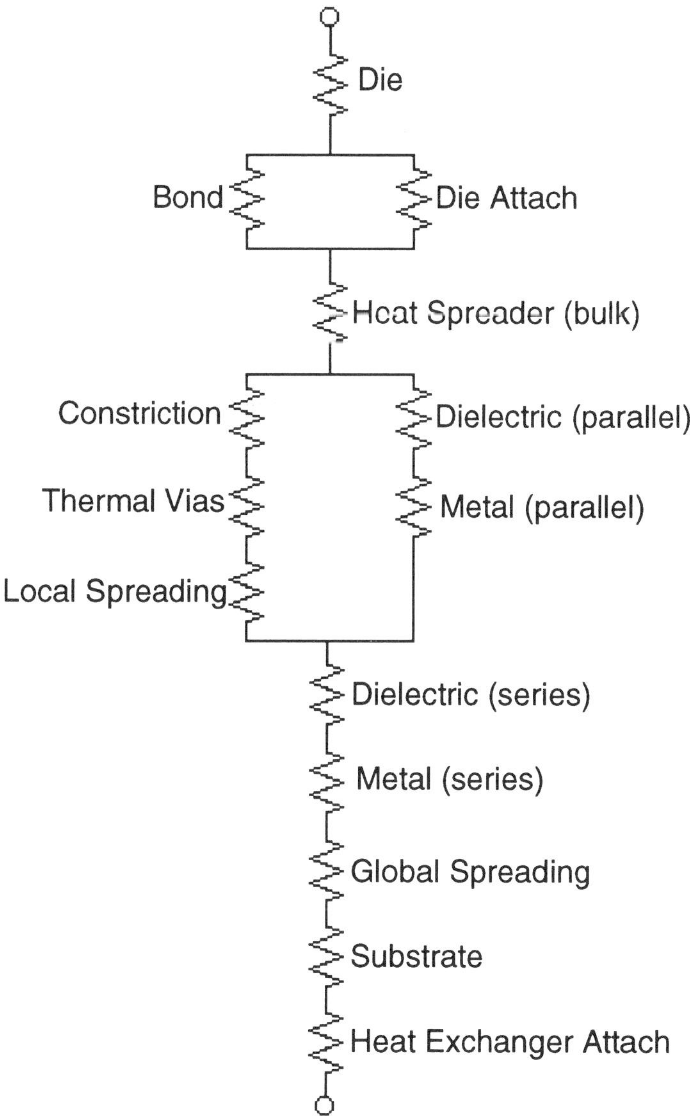

Figure 3.14. The general interconnect/assemble model used for computing the internal thermal resistance for the through-substrate path in a chip-last MCM. If MCM-C or MCM-L are used the substrate component is zero. If thermal vias terminate at the substrate the series metal and dielectric terms are zero. If a face-down bonding approach is used the die term is zero.

In the simplest of models, all series and bulk terms in Figure 3.14 would be computed using a relation similar to (3.33). In reality, however, the heat dissipated by a component does not just flow in paths normal to the die surface, it flows laterally as well ("spreads"). The lateral spreading of heat increases the effective area through which heat is conducted. Many closed form relations have been derived to model spreading and constriction in the thermal path. Different solutions are applicable under different sets of conditions, see [3.18] and the references contained therein for some of the available solutions. To provide a feel for the type of solutions which are needed we have provided the relations for the constriction resistance for heat flux going into thermal vias from either a heat spreader or die attach material.

The resistance derived here could be used for the "constriction" term in Figure 3.14 and is only used if thermal vias are present. Consider the circular portion of the heat spreader associated with a single thermal via shown in Figure 3.15. Solving (3.31) for dT yields,

$$dT = \frac{Q(r)}{k_{hs} 2\pi r t_{hs}} dr \tag{3.34}$$

where,

r = radius

t_{hs} = heat spreader thickness

k_{hs} = thermal conductivity of heat spreader.

Note that the negative sign in (3.31) does not appear in (3.34) because the heat is assumed to flow into the thermal via (Figure 3.15). Since the heat flows into the heat spreader from above, the heat flow (Q) is a function of the radius (r). Q(r) is given by,

$$Q(r) = (\pi r_2^2 - \pi r^2) q \tag{3.35}$$

where q is the heat flux into the heat spreader. Substituting for Q(r) in (3.34) and integrating both sides from R_1 to R_2 the temperature gradient across the heat spreader section is found to be,

$$\Delta T = T_2 - T_1 = \frac{q}{2k_{hs}t_{hs}} \left(r_2^2 \ln \frac{r_2}{r_1} + \frac{r_1^2}{2} - \frac{r_2^2}{2} \right) \tag{3.36}$$

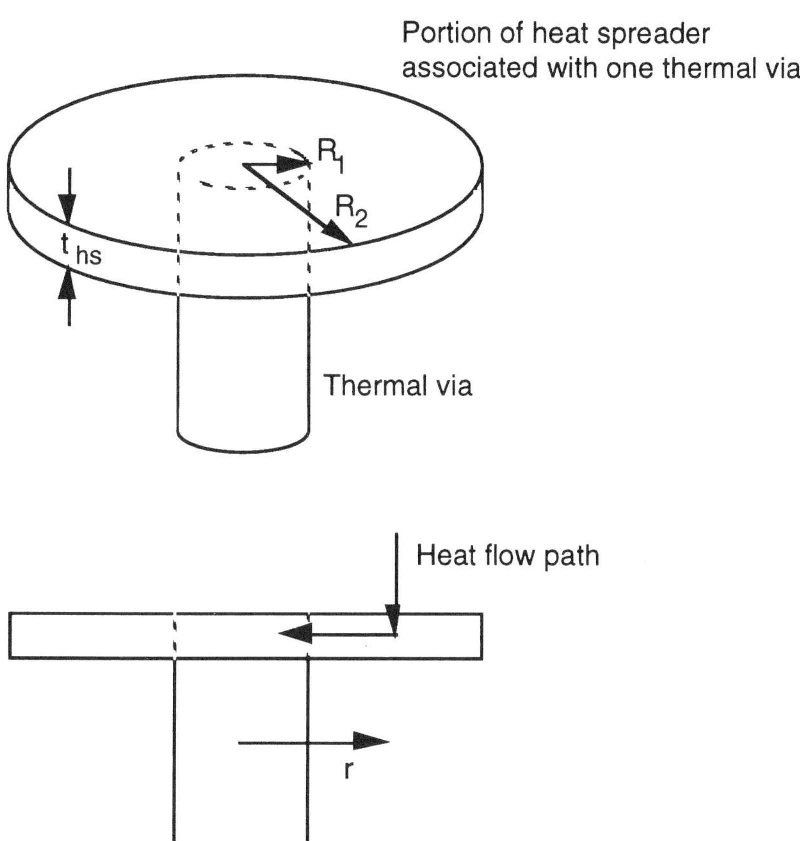

Figure 3.15. Constriction of heat from a thin heat spreader into a thermal via.

Thermal resistance is defined as,

$$R_t = \frac{\Delta T}{\text{area } q} \tag{3.37}$$

therefore, adding the thermal resistances from all the thermal vias under the heat spreader in parallel we obtain

$$R_{\text{heat spreader (constriction)}} = \frac{q}{2k_{hs}t_{hs}\text{area}_{hs}}\left(r_2^2 \ln\frac{r_2}{r_1} + \frac{r_1^2}{2} - \frac{r_2^2}{2}\right) \tag{3.38}$$

where,

$$r_1 = \frac{d_{thermal\ via}}{2}$$

$$r_2 = p_{thermal\ via}\sqrt{\frac{1}{\pi}}$$

$d_{thermal\ via}$ = diameter of the thermal via

$p_{thermal\ via}$ = center-to-center pitch of the thermal vias.

The derivation of (3.38) is only valid under the assumption of "in plane" conduction, i.e., only if the heat spreader is present, thin, and highly thermally conductive, and the material the heat spreader is attached to has a much lower thermal conductivity. If there are no thermal vias then $R_{heat\ spread\ (constriction)} = 0$. If there are thermal vias present but no heat spreader then the constriction resistance from the die attach material into the thermal vias must be computed. The constriction resistance from the die attach material may be estimated using the following relation from Yovanovich [3.19],

$$R_{die\ attach\ (constriction)} = \frac{B}{k_{da}\, n_{thermal\ vias}\, area_{thermal\ via}} \tag{3.39}$$

where,

$B = 0.47890 - 0.62498\varepsilon + 0.11789\varepsilon^3 - 0.000071\varepsilon^5 + 0.02582\varepsilon^7$

k_{da} = thermal conductivity of the die attach material

$$\varepsilon = \frac{\sqrt{\pi}}{2}\frac{d_{thermal\ viaa}}{p_{thermal\ via}}$$

$$area_{thermal\ via} = \frac{\pi d^2_{thermal\ via}}{4}$$

$n_{thermal\ vias}$ = number of thermal vias under the die = $\dfrac{area_{die}}{p^2_{thermal\ via}}$.

External Thermal Resistance. The external thermal resistance (R_{ext}) must be computed considering the type of cooling used for the system, i.e., it is dependencies include the cooling fluid's properties and mass flow rate. The maximum allowable external resistance can be estimated from a thermal budget specified for each component in the system. If the $\Delta T_{coolant}$ term in (3.30) is treated in R_{ext} (see next section), the thermal budget is based on the maximum allowed temperature rise over ambient ($T_{coolant\ in}$) of the component (i.e., T_j - $T_{coolant\ in}$). The type of cooling solution required for a particular module is determined by the component with the smallest allowable external thermal resistance in the system. The limiting external thermal resistance for a component is found from (3.30),

$$R_{ext\ (component)} = \frac{\text{thermal budget} - R_{int} P_c}{P_m} \tag{3.40}$$

where P_c is the power dissipation of the component, R_{int} is the internal thermal resistance associated with the component, and P_m is the total module power dissipation. R_{int} is formed by adding appropriate series and parallel combinations of R_t from (3.29) as shown in Figure 3.14. Fixing the thermal budget at 55 degrees C rise over ambient and plotting the relationship in (3.40) we obtain the relation shown in Figure 3.16.

The generation of a plot like Figure 3.16 for a module design is extremely useful in determining the heat sink requirements on a system. Different interconnect technologies and cooling paths can be compared by locating the internal thermal resistance (multiplied by the component's area) on the horizontal axis, the external thermal resistance requirement is then found from the diagonal line defined by (3.40) corresponding to the component of interest - examples are plotted in Figure 3.16. The goal is to have the heat sink present less external thermal resistance than that given by Figure 3.16 for the component. The external thermal resistance corresponding to several cooling approaches are shown on Figure 3.16.

The external thermal resistance presented by a heat sink is inversely proportional to its area if the chip area to heat sink area ratio is large. However, as the heat sink increases in size its effectiveness will diminish due to spreading resistance. Therefore, an increase in the heat sink area simply shifts the heat sink bands shown in Figure 3.16. In the next

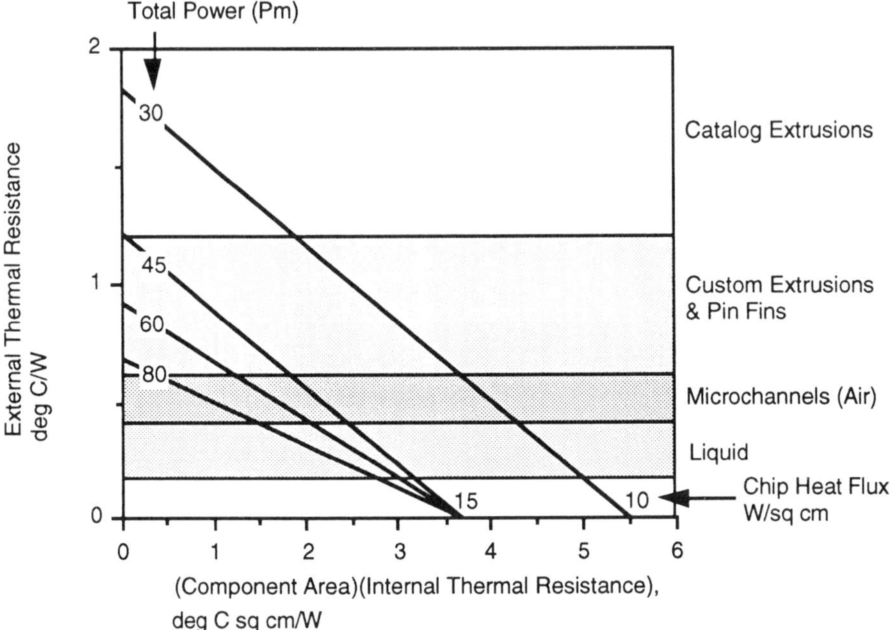

Figure 3.16. Theoretical cooling limits (after [3.20]). The figure assumes a 55 degree thermal budget for each component and the heat sink bands are typical for what could be achieved in a 2.5 x 2.5 x 1 inch heat sink volume. The chip heat flux is given by P_c divided by the component area.

section we discuss how to estimate the external thermal resistance for extended surfaces using air cooling.

External Thermal Resistance Associated with Extended Surfaces. For air cooled systems we wish to estimate the external thermal resistance associated with the extended surfaces (for example, finned heat sinks) which may be attached to chips or modules to remove heat, and determine the constraints on fans which could be used to provide forced convection. The external thermal resistance associated with a fin in the extended surface structure shown in Figure 3.17 is given by,

$$R_{\text{fin-ext}} = R_{\text{fin-convection}} + R_{\text{fin-caloric}} \tag{3.41}$$

The first term in (3.41) is due to convection,

$$R_{\text{fin-convection}} = \frac{1}{\eta h A_f} \tag{3.42}$$

where

η = the fin efficiency
h = the heat transfer coefficient from the surface of the fin
A_f = total fin surface area (= 2ab + at + 2tb for rectangular fins).

The fin efficiency is the ratio of the actual heat transfer from the fin to the heat transfer from the fin if the whole fin were at the base temperature. The fin efficiency accounts for the fact that the fin is not isothermal at the base temperature. For rectangular fins the fin efficiency is given by,

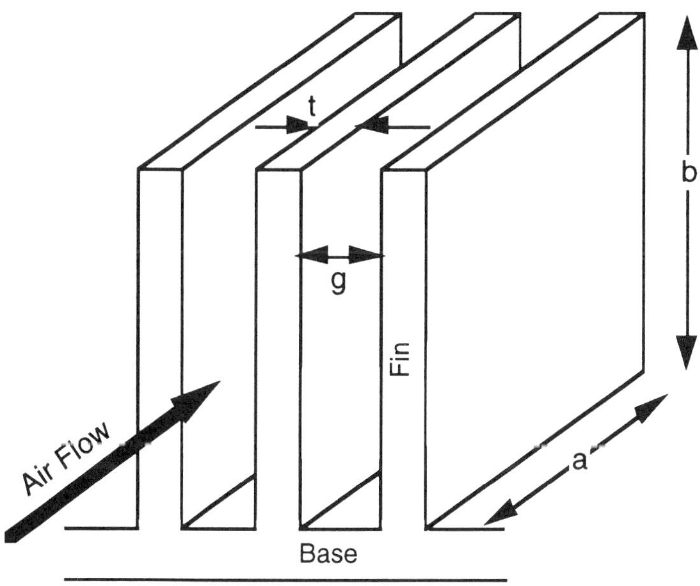

Figure 3.17. Finned extrusions.

$$\eta = \frac{\tanh(ub)}{ub} \tag{3.43}$$

where

$$u = \sqrt{\frac{2h}{kt}} \quad \text{(rectangular fins of thickness t)} \tag{3.44}$$

In (3.44), k is the thermal conductivity of the fin material. Efficiencies for other geometries have been derived, see [3.21]. The fin efficiency accounts for the fact that heat flows from the base to the tip of the fin. If the fin is dissipating heat to the cooling fluid, the fin becomes cooler as the tip is approached. Cooler portions of the fin are less efficient at dissipating power then hotter portions.

The heat transfer coefficient, h, for many different structures has been derived. The heat transfer coefficient is a function of the channel dimensions, thermal conductivity of the cooling fluid (k), and the Nusselt number (Nu),

$$h = \frac{kNu}{D_h} \tag{3.45}$$

where

D_h = the hydraulic diameter of the duct = $\dfrac{2gb}{(g+b)}$

g = width of the duct
b = height of the duct.

The Nusselt number is a function of several parameters including: coolant properties, coolant flow rate, flow regime (laminar, transitional, turbulent), channel geometry, position in the channel, heat input boundary conditions, the type of channel entrance, and the entrance coolant temperature profile. The following assumes that the channel entrance is abrupt, and that the heating rate is constant in the flow direction with uniform peripheral surface temperature. Analytical solutions for the Nusselt number exist for fully developed flow, the thermal entry-length problem, and the combined thermal and hydrodynamic entry-length problem. For heat sinks used in microelectronics applications the flow should be well mixed at the channel entrance and therefore best modeled by the combined flow, however, limited correlations for simultaneously developing flow are available. Therefore, we generally use models for fully developed flows to obtain Nusselt numbers. This approximation leads to lower Nusselt numbers than are actually present and provides a conservative estimate. The Nusselt number as a function of the inverse of the duct aspect ratio for a fully developed laminar-flow is shown in Figure 3.18.

An additional caloric term must be added in (3.41) to account for the heating of the cooling fluid flowing between the fins,

$$R_{\text{fin-caloric}} = \frac{1}{\rho c_p A_d V} \tag{3.46}$$

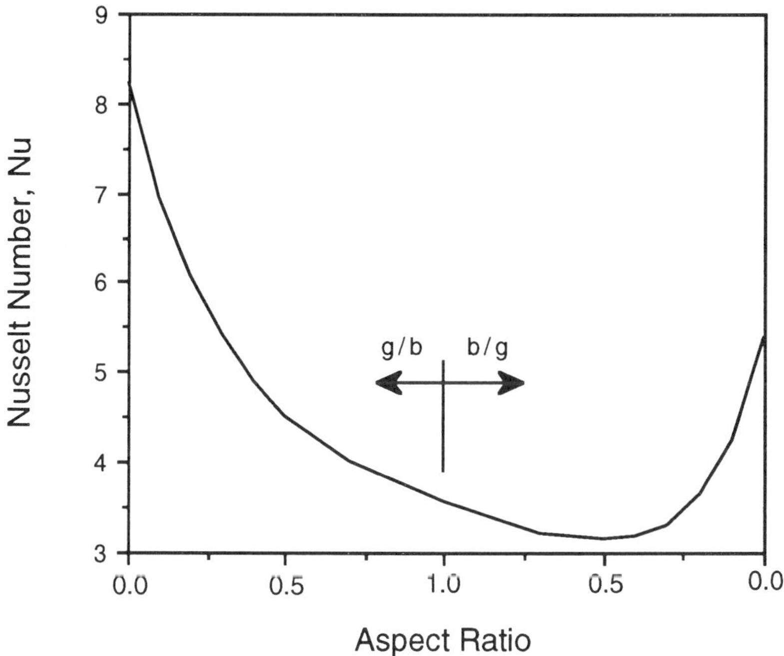

Figure 3.18. Nusselt number of a fully developed laminar flow in a rectangular duct, [3.22].

where,

ρ = mass density of the cooling fluid
c_p = specific heat of the cooling fluid
A_d = cross-sectional area of the duct (gb)
V = velocity of the cooling fluid.

Some references include the fin efficiency in the denominator of (3.46), [3.23]; a discussion of the inclusion of the fin efficiency in (3.46) appears in [3.22] (page 90).

The properties which describe a cooling fluid are dependent on the pressure, temperature, and many other environmental properties. For a given fluid the values of the describing properties have been tabulated into handbooks and will not be repeated here. The characteristics of air a various temperatures appear in [3.24]. One remaining problem with the present analysis is to determine the temperature of the cooling fluid. The change in the coolant temperature can be expressed as,

$$\Delta T_{coolant} = (T_{out} - T_{in}) = \frac{Q}{mc_p} \tag{3.47}$$

where m is the coolant mass flow rate (ρVA) and Q is set equal to the module power dissipation, P_m. The solution of (3.47) is iterative since ρ and c_p are both functions of the temperature of the coolant. To first approximation, the average temperature of the coolant is set equal to $(T_{out} - T_{in})/2$.

The analysis above (3.41)-(3.47) allows the estimation of the external thermal resistance corresponding to a single rectangular fin. If the heat flow into the heat sink is approximately homogeneous across the whole structure, then the external thermal resistance is the parallel combination of the external thermal resistances all the fins or pins.

Figure 3.19 shows a plot of the external thermal resistance associated with heat sinks having rectangular fins.

Fan Selection for Forced Air Systems. For electronic systems cooled by forced convection, the pressure drop characteristics of the system must be determined in order to select an adequate fan or blower. To determine the fan characteristics, the minimum allowable flow rate for which none of the system components will exceed their maximum al-

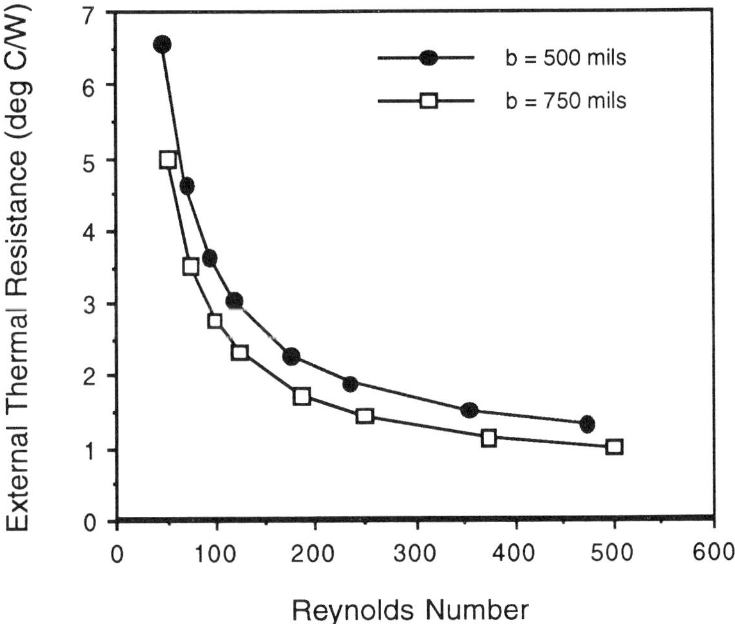

Figure 3.19. External thermal resistance associated with heat sinks having rectangular fins. The aluminum heat sink size is fixed at 2.5 inches square and the cooling fluid is assumed to be air with an inlet temperature of 273 K. The fin separation is 95 mils and the fin thickness is 5 mils. The Reynolds number is given by VD_h/υ where υ is the cooling fluid viscosity.

lowable junction temperatures must be determined. An air cooled system, pressurized by a fan, has a finite possible air flow volume for a given static pressure drop. Figure 3.20 shows the static pressure curve for a fan and an electronic system which it could be used to cool. In practice the fan curve can be obtained from vendor performance specifications and the system flow loss curve would be obtained through experimentation. The point at which the two curves intersect is the system operating point, i.e., if this fan is used in this system this is the air flow and pressure which will result. If the air flow and pressure necessary to cool the system falls inside the rectangle in Figure 3.20 then the fan is adequate to meet the cooling requirements.

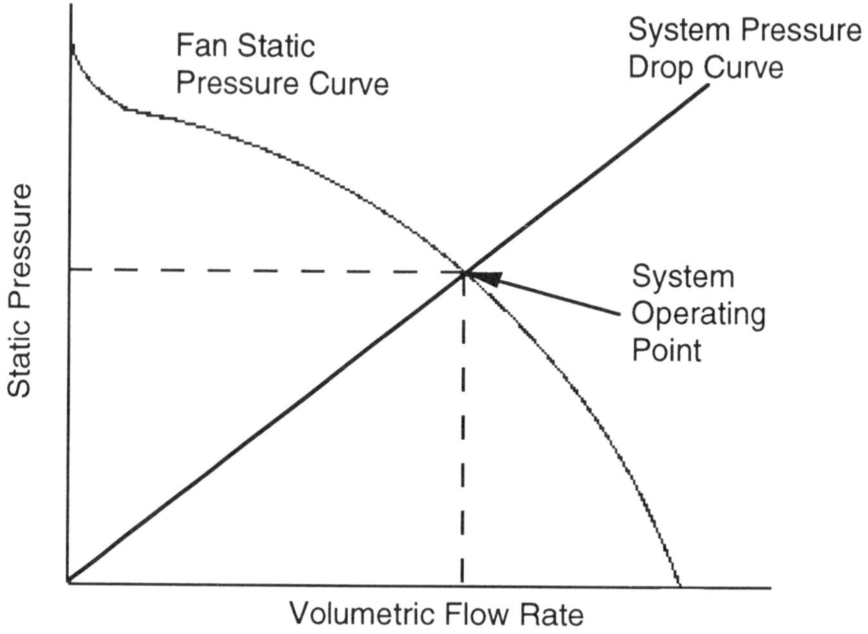

Figure 3.20. Static pressure curve for a commercially available muffin fan. Volumetric air flow rate is the volume of cooling fluid moved through the system per unit time.

Each cooling fluid flow channel in a system has an associated pressure drop. The pressure drop is related to the physical properties of the cooling fluid, the flow velocity, the surface characteristics of the duct, and the dimensions of the space through which the fluid is flowing. For laminar flow through straight ducts[3],

$$\Delta P = \frac{f \rho a V^2}{2 D_h} \tag{3.48}$$

3. Sometimes (3.48) is written with the acceleration of gravity (g) in the denominator. The form which includes g is only applicable if English units are used, i.e., g implicitly converts lbs-mass from the density to lbs-force.

where,

$$f = \text{friction factor} = \frac{64}{Re}$$

$$Re = \text{Reynolds number} = \frac{VD_h}{\upsilon}$$

υ = kinematic viscosity of the cooling fluid
ρ = mass density of the cooling fluid
a = length of the duct in the direction of the cooling fluid flow (shown in Figure 3.17)
V = velocity of the cooling fluid.

For Reynolds numbers less than $2x10^3$ the flow in pipes is laminar while above $2x10^3$ the flow transitions into turbulence. The friction factor (f) depends on the surface roughness, flow regime, and the Reynolds number [3.25]. Equation (3.48) assumes that the flow path is straight and has a constant area. If the flow path turns or changes sizes, alternative formulations can be used (see [3.25]). The various pressure drops can be accumulated in series and parallel similar to electrical and thermal resistances in order to determine the system flow impedance. The system pressure drop curve in Figure 3.20 is found by plotting the volumetric flow rate (m/ρ) versus the static pressure (ΔP). Using (3.48) the pressure drops for various extrusions are plotted in Figure 3.21.

The analysis above will provide best case estimates for most heat sinks since it assumes that all the air impinging on a heat sink goes through the structure and not around it.

Acoustic Noise. Acoustic noise is a serious concern associated with forced air cooling systems, especially workstation and personal computers. Acoustic noise is a difficult parameter to estimate for air cooled heat sinks because noise generation is associated with the non-conservative details of the flow in the heat sink region. A simple procedure for estimating the acoustic noise is to compute the pumping power for a flow. The extrusion analysis discussed above produces estimates of pressure drop, and the flow was assumed to be ducted, so the pumping power can be computed from,

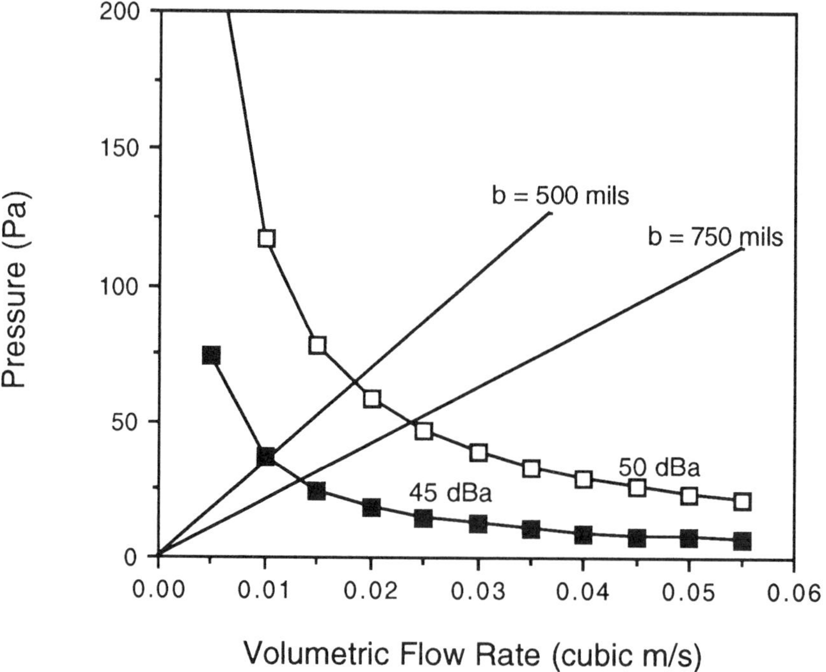

Figure 3.21. Pressure drop associated with the extrusions considered in Figure 3.19. This figure assumes that all the air flow goes through the ducts in the heat sink. The volumetric flow rate is the product of the air velocity and the cross-sectional area of the duct multiplied by the number of ducts. The 45 dBa and 50 dBa curves represent acoustic noise and are explained in the next section.

$$\text{Pumping Power} = \frac{mn}{\rho}\Delta P \tag{3.49}$$

where n is the number of identical ducts in the heat sink. This is the power loss associated with moving air through the heat sink, some fraction of which shows up as acoustic noise power. By comparing the value computed using (3.49) to acoustic noise estimates for commercial fans and cabinet blowers, the average acoustic noise power is estimated to be approximately,

$$\text{Noise Power} = 4 \times 10^{-9}\,\text{Pumping Power} \tag{3.50}$$

Curves corresponding to two different noise powers are included in Figure 3.21[4]. Note that (3.50) is only a rough order of magnitude estimate of the noise power (acoustic noise really represents a system level problem that may not be accurately treated by just considering a single heat sink).

---

4. The noise power in dBa is given by $10\log_{10}\left[\frac{(\text{Noise Power (W)})}{P_o}\right]$ where P_o is a reference power (10^{-12} W).

3.3.4 Electrical Analysis

Communicating information from one chip to another within a module requires the signal level on a transmission line to change. The transition to a new signal level does not occur instantaneously due to propagation delay and the non-zero transition times of drivers on chips. Decreasing signal transition times facilitates better performance (i.e., faster computers), but also magnifies a variety of phenomena that are undesirable and often detrimental to a module performing its intended task, namely noise.

Three electrical parameters - resistance, capacitance, and inductance are always present in packaging and interconnect systems. Resistances cause dc drops on signal lines and contribute to RC charging delays, both of which are unwanted effects. Resistances, however, can help damp unwanted noise in systems. Resistance in signal lines increases as line geometries shrink and can be several ohms per centimeter in thin film MCM interconnects. For very fast systems, resistance is also a function of frequency due to skin effects. Capacitance is the primary cause of delays in packaging and interconnect systems. Capacitive loading on signal lines is decreased by using bare die instead of die in single chip packages and by reducing net lengths. Capacitive coupling between power and ground in power distribution systems is beneficial for managing switching noise. Capacitance and resistance can be either lumped (discrete load or termination) or distributed (such as the resistance and capacitance associated with a transmission line). Inductance contributes to switching noise and time-of-flight delays in packaging systems. Inductance can be reduced by shortening the length of electrical connections in modules (such as bond wires or connector traces) or by replacing bus structures with planes. The effective inductance in modules is also reduced by increasing the circuit return path mutual inductance which can be accomplished by decreasing the pitch between bonding leads. Capacitive and inductive coupling between adjacent signal lines also causes crosstalk.

The basic elements of interest in electrical performance are delays and noise. These elements are characterized through the measurement of crosstalk, switching noise, dc drops, attenuations, and critical path delays.

Electrical Analysis Hierarchy. Application specific electrical analysis is not straightforward at the conceptual design level. Unlike thermal or size analysis, estimation level models which capture the application specific noise or delay properties are limited. Specific information generally requires the use of simulation programs (at the very least circuit simulators). A hierarchy of electrical analysis methods can be defined:

Figure of Merit - Simple electrical packaging figures of merit are easily computable. These estimates are technology dependent but application independent. They are useful for comparing the general merits of one interconnect or bonding technology over another but do not provide any application specific performance detail. Examples include the calculation of saturated near-end crosstalk or power-delay product.

Closed-Form Design-Dependent Models - These models capture simple design specific details such as the number and size of loads on a transmission line. Analysis in this case is limited to the formulation of closed-form expressions. This type of modeling is practical at the conceptual design level, however, it usually results in performance measures which are acceptable for relative comparison of one technology/architecture combination to another, but not acceptable on an absolute accuracy basis. An example is the estimation of a critical net delay by adding closed-form expressions for the RC charging delay and time-of-flight delays. Such a summation is useful for comparison purposes but does not model real nets accurately since delays on most real nets are dependent on reflections, crosstalk, and switching noise.

Simulation Models - Full scale simulation of packaging structures is the only method which allows electrical performance predictions with measurable absolute accuracy to be obtained for real structures. Accurate results can be obtained from closed-form expressions only under very special circumstances which generally do not represent realistic situations. The drawback of using simulation models at the conceptual design level is the computational and problem setup overhead involved (see the estimation-simulation gap discussion in Chapter 1). An alternative, which is practical at the conceptual design level if a CAD system for conceptual design is used, is the managed use of simple simulation coupled with estimation [3.26].

A further complication in electrical analysis is the necessity that the analysis treat the whole system in order to be useful in optimizing the design of a module. Unlike other views of the tradeoff space, in electrical analysis the chips can not be decoupled from the packaging, i.e., for truly concurrent design, drivers and receivers would be optimized for a particular combination of packaging technology. Comparing electrical performance by fixing driver and receiver characteristics and varying the packaging technologies does not represent an "apples-to-apples" comparison if the chip characteristics are under the control of the designer. Similarly, switching noise analysis requires modeling of the power and ground distribution system external to the module under design.

An excellent treatment of electrical modeling of packaging and interconnect structures is contained in [3.9] and readers are encouraged to refer to this reference for additional information on electrical analysis approaches. This section focuses on the formulation of figures of merit and closed-form approaches to electrical performance evaluation.

Delay Analysis. Probably the most important area in electrical performance for potential comparison between different packaging combinations is delay. Packaging delays, along with the delays associated with gates, receivers, and memory access times will determine the maximum operating frequency of multichip systems. The packaging delay not only depends on the length of nets, but also the distribution of loads on the nets, the size of the loads, and the characteristics of the circuits driving the nets.

The total delay in a circuit path can be approximately decomposed into the following elements:

- RC charging delays
- Delays associated with noise (crosstalk, switching noise, reflections)
- Time-of-flight delays ($\sqrt{LC}$ delays)
- Delays associated with driver and receiver circuitry

- Delays associated with memory access (if applicable)

In a digital system the voltage level at the interest to receiving circuitry determines the definition of delay. In static logic circuits it is customary to measure delay from the 50% level of the input signal to the 50% signal level at a receiver of interest. In precharged circuits, and for certain critical memory paths, it may be necessary to consider the 90% signal level in order to guarantee that transistor switches are fully turned on or off. The important signal level depends on the threshold of the receiver circuitry.

RC Charging Delays. The simplest delay estimation is the lumped RC or Elmore delay model [3.27]. In the lumped RC model the delay is computed as the product of lumping all the resistances and all the capacitances together along a path,

$$t_{charging} = \sum_i R_i \sum_j C_j \tag{3.51}$$

The lumped RC model is often used to estimate on-chip delays where interconnect distances are small and time-of-flight delays are negligible compared to RC charging delays.

The lumped RC model can be improved upon by considering the resistances and capacitances as distributed along the interconnect. For the simple circuits shown in Figure 3.22, a charging delay calculation is formulated as

$$\begin{aligned} t_{charging} = {} & \alpha_1 R_{on}(C_1 + C_2 + C_3 + C_4 + C_5) + \alpha_2 R_{on} C_2 \\ & + \alpha_1 R_2 (C_3 + C_4 + C_5) + \alpha_2 R_2 C_4 + \alpha_1 R_2 C_5 \end{aligned} \tag{3.52}$$

where R_{on} is the on-resistance of the output stage and, α_1 and α_2 are coefficients which depend on the signal level to which one wishes to measure the delay.

Equation (3.52) is often used to model on-chip connections where the interconnect lines may be accurately approximated as distributed RCs. In this case, R>>ωL for the signal interconnects over the entire

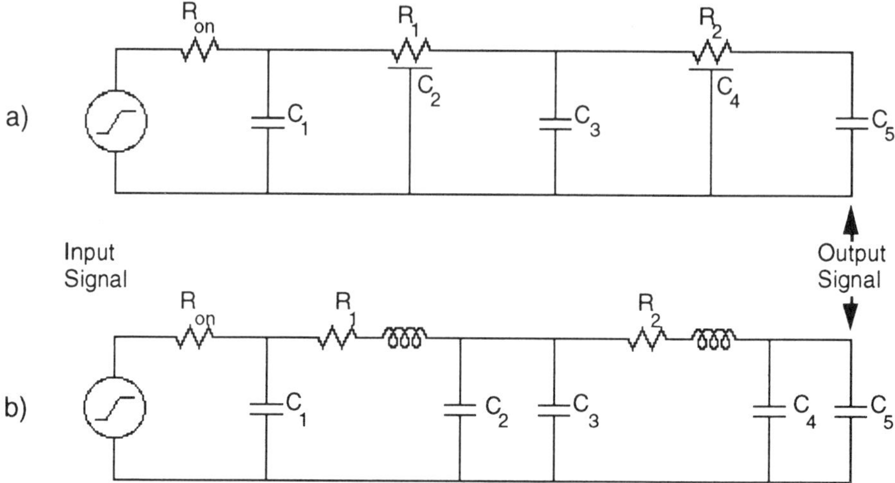

Figure 3.22. Simple equivalent circuit representations of loaded nets. a) Distributed RC model, b) Lumped RLC model, R_1 and R_2 represent the loss in the two transmission line segments.

frequency spectrum of the signal and the series resistance of the interconnects is often greater than the characteristic impedance of the line (ω is the frequency in radians). Under these circumstances (3.52) represents the entire delay, i.e., if the inductance is negligible, there is negligible time-of-flight delay. The model shown in Figure 3.22b corresponds to the case where connections are assumed to be lossy transmission lines and inductances are not negligible. In this case, delays associated with the inductance of the transmission line (i.e., time-of-flight delay) must be considered in addition to (3.52) in order to approximate the total delay of the line.

In (3.52), α_1 is used for a lumped RC network and α_2 is used when a uniformly distributed RC network is required. If the lumped model in Figure 3.22b is used, α_2 is replaced by α_1 in (3.52). Finding α_1 and α_2 at a given signal level for a step input (signal line rise time, $t_r = 0$) is straightforward. The voltage across the capacitor in a parallel RC circuit is given by

$$V_c(t) = V(1 - e^{-t/RC}) \tag{3.53}$$

where V is the logic level of the signal line driver. For example $V_c(t)$

reaches 90% of V when t = -RC ln(1/10) so α_1 is equal to -ln(1/10) = 2.3. Solving for α_2 is more complex. In order to find α_2, (3.54) must be solved for t, [3.28],

$$V_c(t) = V\mathrm{erfc}\sqrt{\frac{RC}{4t}}\ , t \ll RC \tag{3.54a}$$

$$V_c(t) = V\left(1 - 1.366e^{-2.5359t/RC} + 0.366e^{-9.4641t/RC}\right), t \gg RC \tag{3.54b}$$

For the 50% signal level $\alpha_2 = 0.4$ and for the 90% level $\alpha_2 = 1.0$.

Time-of-Flight Delays. In this discussion, we define the time-of-flight delay as the delay governed by the speed of light in the dielectric media surrounding the signal line, i.e., the $\sqrt{LC}$ delay. To obtain time-of-flight delays, a simple unloaded propagation speed calculation can be used

$$t_{tof} = \mathrm{length}\sqrt{\varepsilon_o \varepsilon_{eff} \mu_o} \tag{3.55}$$

where ε_o and μ_o are the permittivity and permeability of free space, and ε_{eff} is the effective dielectric constant of the material embedding the line. The effective dielectric constant is not only a function of the dielectric material but can also be a function of the line geometry and construction (see [3.29] for examples of ε_{eff} calculations). Many references formulate a modified version of (3.55) for transmission lines with distributed capacitive loads,

$$t'_{tof} = t_{tof}\sqrt{1 + \frac{C_L}{C_o}} \tag{3.56}$$

where t_{tof} is given by (3.55), C_L is a load capacitance distributed uniformly over the line, and C_o is the characteristic capacitance of the line. Equation (3.56) approximates a loaded transmission line as a new transmission linc characterized by t'_{tof} and $Z'_o = t'_{tof} / (C_o + C_L)$.

The total delay on a loaded transmission line is some combination of RC charging delays and time-of-flight delays (plus other effects due to noise and reflections). Often the delay is estimated using either (3.52) alone (for low speed systems), or, (3.55) or (3.56) alone (for high speed systems). For situations in which neither RC charging nor inductive effects are negligible (lossy transmission lines for example), summing the two delay components may be an acceptable approximation for tradeoff purposes. One approximation of the total delay of a packaging and interconnect system can be found by summing the RC charging delay (3.52) and the time-of-flight delay from (3.55). An alternative to this is to add the charging delays as given by (3.52) without including delays due to capacitive loads distributed along the transmission line (C_1 and C_3) to a sum of the time-of-flight delays given by (3.56). Both of these methods result in approximately the same total packaging delay. However, the first method gives a more accurate description of the separate delay components.

RC versus Time-of-Flight Delays. Figure 3.23 shows the relative sizes of the RC charging delay and the time-of-flight delays for two different interconnect technologies. For this simple comparison, a point-to-point net was chosen with a load capacitance of 3 pF at the end, and a driver with an on-resistance of 25 Ω. In both cases, the line was assumed to be a 50 Ω copper stripline and delay was measured to the 50% level.

For short lines, both cases are dominated by RC delays (due to the on-resistance of the driver), as the line length increases the contribution from time-of-flight increases. In both cases, the time-of-flight delay becomes dominant for line lengths in excess of 2 cm. In the thin film interconnect case, the RC charging delay becomes dominant again for very long lines due to the large line resistance (2.4 Ω/cm).

Although the printed circuit board (PCB) and thin film interconnect may have similar delays for similar line lengths, it is not necessarily valid to conclude that the same application implemented on a printed circuit board will have the same electrical performance as its thin film multichip module (MCM) implementation. Thin film MCMs provide the opportunity to place components closer together and therefore line lengths are often shorter. If single chip packaged die are used on printed circuit boards, the capacitive loading on the line may also be greater, contributing additional RC delay.

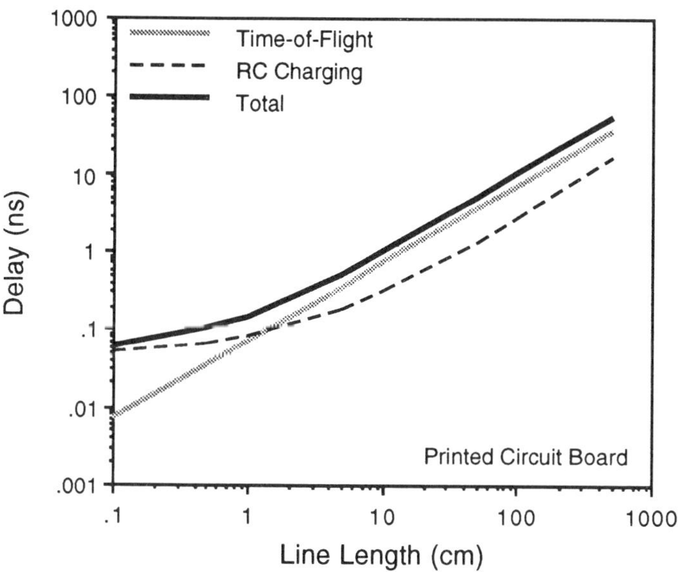

Figure 3.23a. Printed circuit board (PCB) delay components. An FR-4 board was assumed ($\varepsilon_{eff} = 4.5$) with a signal line width of 8 mils and thickness of 1.4 mils.

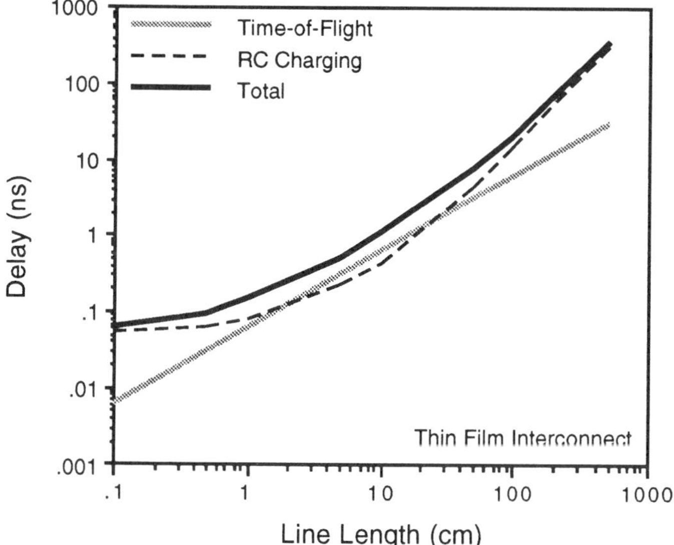

Figure 3.23b. Lossy thin film interconnect delay components. An polyimide dielectric was assumed ($\varepsilon_{eff} = 3.4$) with a signal line width of 15 μm and thickness of 5 μm.

Circuit Accessibility. The simplest delay "figure of merit" is a measure of circuit accessibility [3.30]. Circuit accessibility measures the number of circuits that can communicate with each other within a specified period of time (usually 1 ns), Figure 3.24. This metric asserts that packing the greatest number of circuits into the smallest possible area is a valid predictor of the minimum cycle time for a given chip/packaging technology combination.

A simple expression for chip-to-chip delay derived by summing (3.52) and (3.55) is given by,

$$t_d = 0.69R_{on}(C_o\mathcal{L} + C_L) + 0.4R_oC_o\mathcal{L}^2 + 0.69R_o\mathcal{L}C_L + \mathcal{L}\sqrt{\varepsilon_o\varepsilon_{eff}\mu_o} \tag{3.57}$$

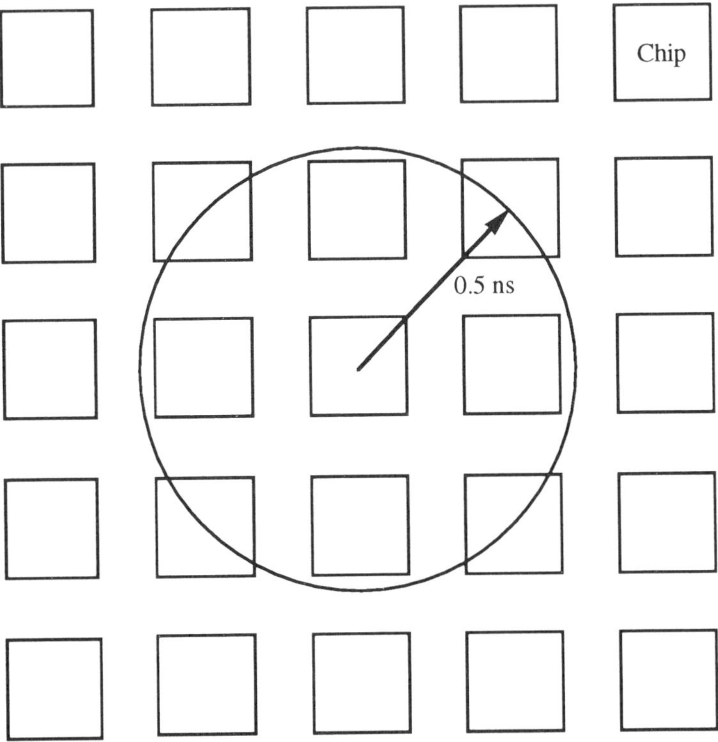

Figure 3.24. Definition of circuit accessibility (circuits/nanosecond).

where,

R_{on} = the on (or output) impedance of an off-chip driver
C_o = capacitance per unit length of a signal line
R_o = resistance per unit length of a signal line
C_L = load capacitance of an on-chip receiver
$\mathcal{L}$ = the distance
ε_o = permittivity of free space ($8.85\text{x}10^{-12}$ F/m)
ε_{eff} = dielectric constant
μ_o = permeability of free space ($4\pi\text{x}10^{-7}$ H/m).

Equation (3.57) is a simple sum of the RC charging delay and time-of-flight delay for a lossy net with one load on it. To compute the circuits per nanosecond delay metric, t_d, is set to 0.5 ns and (3.57) is solved for $\mathcal{L}$, the metric is then given by,

$$\text{circuits/nanosecond} = \frac{\pi\mathcal{L}^2}{4F_p^2}N_{cir} \quad (3.58)$$

where,

F_p = chip footprint dimension (as computed in Section 3.3.2)
N_{cir} = number of circuits per chip (could be computed using Rent's rule, (3.1)).

The model represented by (3.57) and (3.58) is a good example of a figure of merit analysis method. This model contains no net specific details, only information describing the interconnect technology is relevant, i.e., it assumes that all nets have a fanout of one and are point-to-point wired in a straight line from the driver to the receiver. It is also idealized in the sense that it assumes reflections and noise do not affect the delay.

Figure 3.25 shows the application of (3.57) and (3.58). The example shown in Figure 3.25 was formulated by solving (3.57) and (3.58) simultaneously for the CMOS gate arrays described in [3.30].

Estimating the total real delay of a transmission structure is not a trivial exercise. The simple addition of the RC charging delay and time-of-flight delays for a net provides a generic comparison of the packag-

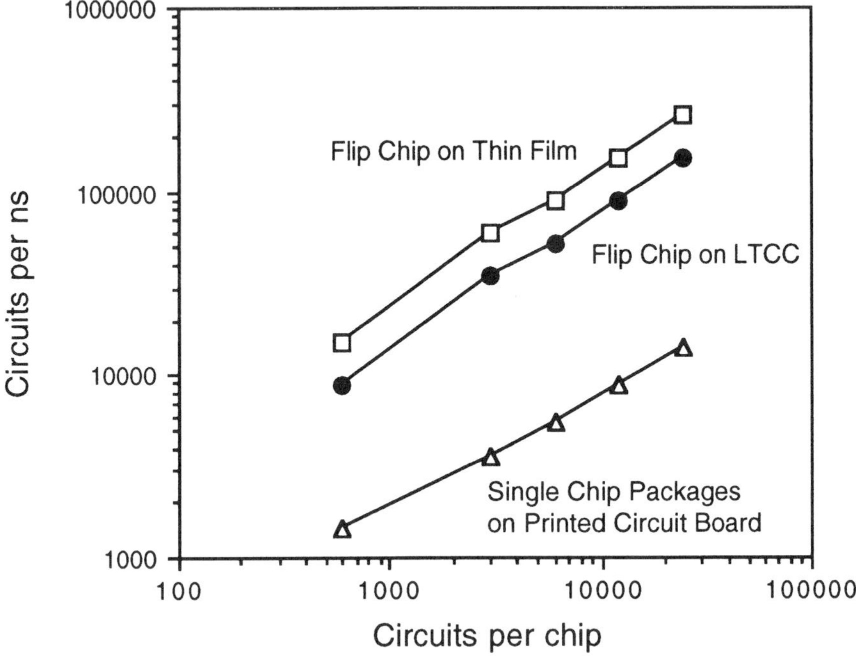

Figure 3.25. Circuit accessibility for surface mount PCB, and flip chip assembled thin film, and LTCC (Low Temperature Cofired Ceramic) modules. The design rules for PCB and thin film signal lines are the same as those used in Figure 3.23. For LTCC the dielectric was assumed to have a dielectric constant of 9.5, signal line width = 4 mils, and signal line thickness = 0.7 mils. The driver on-resistance was assumed to be 25 Ω and the load capacitance was set at 3 pF for the thin film and LTCC cases, and 5 pF for the surface mount/PCB case.

ing technologies but does not accurately represent the real delay in a path. The actual transmission delay in a critical path is also dependent on reflections and noise. The effect of reflections on delay has been treated analytically [3.31], but is most easily accounted for by formulating equivalent circuit models of transmission structures and using a circuit simulator. The inclusion of noise effects on delay is not generally possible at the conceptual design level. While the estimation of worst case noise magnitudes is possible, the timing of the noise disturbances is critical to their effect on the delay. The inclusion of noise effects on delays usually requires simulation of the interconnect system which requires knowledge of wiring adjacent to the critical path, and timing information.

Critical Path Models. Thus far, we have only estimated the delays associated with loaded transmission structures. The important parameter in the design of multichip systems is the delay associated with critical paths which include transmission structures as well as gate delays and possibly memory access times.

The estimation of critical path delays (which may be a system cycle time) is not a trivial task. The cycle time for a central processor is determined by the logical functions that need to be performed in each machine cycle. These logical functions, historically, have used as many as 10-20 logical stages, where each stage is a logic gate. Depending on the semiconductor, packaging, and level of integration used, a logical function may require information to be processed through 5-15 different chips, and 3-5 package entities to complete the function, [3.32].

Every time a logic function is performed over several chips or several blocks within a chip, there is a delay penalty paid for crossing from one chip to another or from one block to another. This penalty is called chip crossing delay. The chip crossing delay is a function of the fanout of the critical net, the driver and receiver designs, and the packaging and interconnect environment. To compute the chip crossing delay, the number of crossings has to be related to the number of entities (blocks, chips, module, boards, etc.). Balderes and White [3.32] provide the relationship shown in Figure 3.26.

Figure 3.26 resulted from an analysis of design statistics from existing machines and VLSI chips, however, these results may be applied to any level of packaging. The information in Figure 3.26 may be used to generate relative packaging delay information. Figure 3.26 provides an understanding of the entity capacity at each level of the packaging separated by a physical packaging boundary. The average distance be tween entities (pitch) determines the delays,

$$t_d = \sum_{i=1}^{N_L} (NL_{avg} t_{\text{delay/length}}) \tag{3.59}$$

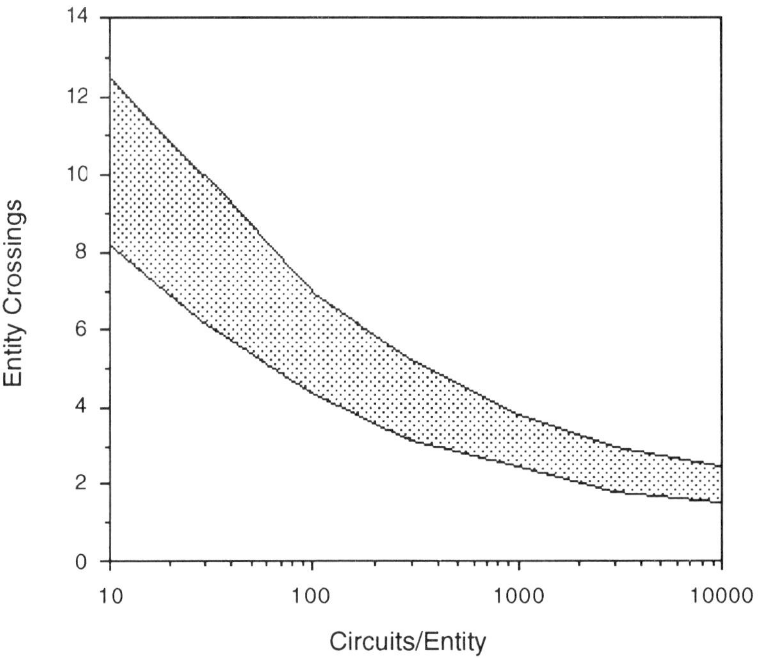

Figure 3.26. Number of entity crossings in a typical logic path as a function of the entity density, [3.32]. (© 1985 IEEE)

where,

N = number of crossings (determined from Figure 3.26)

N_L = number of packaging levels

L_{avg} = average net length at the packaging level (could be computed using R_mP where R_m is determined from equations (3.9) or (3.10) in Section 3.3.1, P is the entity center-to-center pitch, and f is the net fanout)

$t_{delay/length}$ = delay per unit length (time-of-flight delay per unit length for packaging levels, RC estimated delay for chip).

As an example of the use of (3.59), consider the packaging level information presented in Table 3.2. Using this information in (3.9), (3.10), (3.55), and (3.59) allows the estimation of the delay associated with each packaging level. The example in Table 3.2 is a system consisting of 10 cards mounted on a backplane, each card has 100 chips on it. The chips are modeled after series gated ECL chips.

Table 3.2. Crossing delay calculation example.

	1	2	3
Number of "circuits/entity"	500 circuits/ chip	100 chips/ card	10 cards/ board
Number of crossings (from Figure 3.26), N	4.5	7.0	12.5
Rent's constant, p	0.65	0.6	0.6
Average interconnect length from (3.9) or (3.10), R_m (units = number of chip footprints)	4.15	2.80	1.67
Entity center-to-center pitch, P	270 μm	3.0 cm	2.0 cm
relative dielectric constant, ε_r	11.7 Si	4.5 FR-4	4.5 FR-4
delay per unit length, $t_{tof}/\mathcal{L}$	0.4 ns/cm	0.071 ns/cm	0.071 ns/cm
Delay	200 ps	4.2 ns	3.0 ns

The estimates in Table 3.2 do not consider multiple fanout nets. Chip crossing delays have been used to compare packaging technologies, see [3.33, 3.34].

Several models have been used to estimate critical path delays. All critical path delay models involve the summation of a string of delays representing the critical signal path. The most straightforward approach for comparing technologies (chip and packaging) is to define a common function and sample architecture, then sum all the delays in a cycle. The following metrics have been suggested by Gheewala [3.35]. For a scientific computer consider the addition of two large numbers and the following sample architecture: Carry Look Ahead/Carry Save Adder for an ALU,

$$t_{cycle} = \text{6 gate delays} + \text{N chip crossing delays} + \text{latch} + \text{clock skew} \quad (3.60)$$

where N is the number of chip crossings (discussed above). For a large mainframe computer the cache access time is often the limiting process in determination of the performance,

$$t_{cycle} = \text{10 gate delays} + \text{memory access} + \text{3 chip crossing delays}$$
$$+ \text{1 card crossing} + \text{latch} + \text{clock skew} \quad (3.61)$$

VLSI microcomputers rely on compressing the entire processor into a single chip (CPU). The cycle time in these cases can be computed from

$$t_{cycle} = \text{20 gate delays} + \text{register file access} + \text{latch} + \text{clock skew} \quad (3.62)$$

Equation (3.62) does not include any chip crossing delay because the entire instruction path is within one chip.

Another example of this type of delay metric was developed in [3.36]. The figure of merit is given by,

$$t_{cycle} = \text{9 gate delays} + \text{3 buffer delays} + \text{minimum path delay}$$
$$+ \text{average path delay} + \text{maximum path delay} \quad (3.63)$$

This delay metric is truly a "figure of merit" calculation since it does not correspond to the approximation of any real physical delay in a system.

Another figure of merit metric for a critical path is the weighted gate delay. Weighted gate delays are formed by summing the gate, driver/receiver, package, and interconnection delays as above and dividing by the total gate count [3.37]. The weighted gate delay varies linearly with the actual cycle time of real systems. Figure 3.27 shows a weighted gate delay versus relative cycle time for a set of DEC and IBM computer systems.

One possible formulation of a weighted gate delay is given in (3.64),

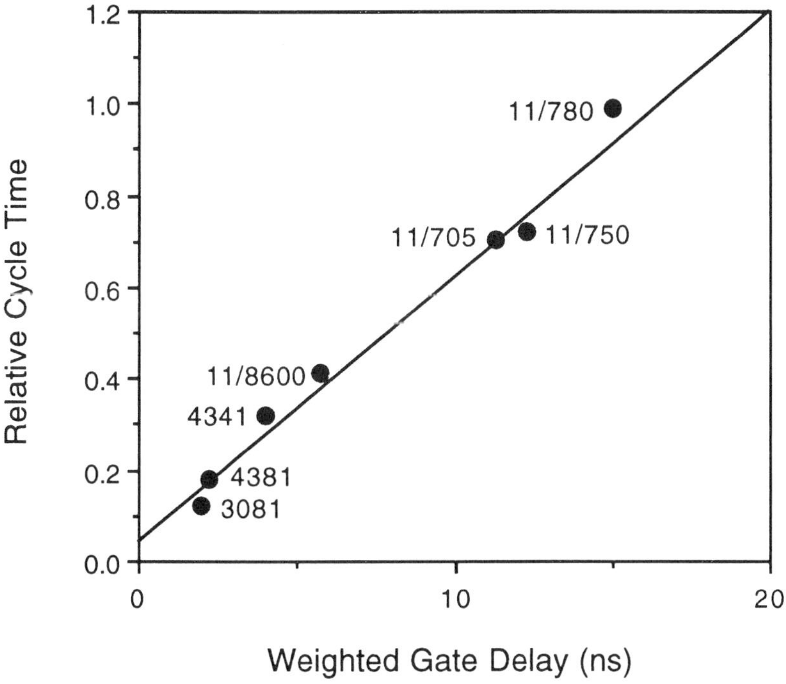

Figure 3.27. Weighted gate delay versus relative cycle time for a group of IBM and DEC computer systems, [3.37].

$$\tau_{wgd} = \frac{1}{k}\left(D_0 + \frac{D_1\,(1 - Q_1^{-0.5})}{E_1^{0.25}} + \frac{D_2\,(1 - Q_2^{-0.5})}{E_2^{0.25}}\right) \qquad (3.64)$$

where,

k = gate depth of a circuit (number of circuits per gate)
D_0 = gate delay
D_1 = chip-to-chip delay
D_2 = module-to-module delay
Q_1 = number of chips
Q_2 = number of modules
E_1 = circuits per chip
E_2 = circuits per module.

More detailed critical path delay estimations can be achieved by including the characteristics of the gates themselves in the delay calculations. Numerous authors have performed a variety of estimations which include the details of gate fabrication and design, [3.9], [3.38], and many others.

Power-Delay Product. Power-Delay Product is a useful and often--quoted metric for chips. The power-delay product is the product of the average power consumption and the average gate delay and can be interpreted as the energy consumed for each logic decision made. Although the power-delay product is usually applied to the comparison of logic families (which is outside the scope of this book), it can be used to predict optimums in some packaging parameters.

Assuming that the value of the power-delay product is approximately independent of design rule changes within a technology family, i.e., for a given type of logic (CMOS or ECL for example) the power-delay product is approximately independent of what design rules are used. Following the derivation of Keyes [3.39], the power-delay product for a logic circuit is given by,

$$U = p_c t_c \tag{3.65}$$

where p_c and t_c are the power dissipation and the delay per circuit. If the maximum heat, which can be removed from a unit area is given by q_A then the following inequality must hold in order to successfully package the chip,

$$q_A F_p^2 \geq p_c N_{cir} = P_c \tag{3.66}$$

where F_p is the footprint dimension of the chip and N_{cir} is the number of circuits per chip. The packaging delay is approximated by

$$t_p = t_c + t_{RC} + R_m F_p \sqrt{\varepsilon_o \varepsilon_{eff} \mu_o} \tag{3.67}$$

where t_{RC} is the RC charging component of the packaging delay and R_m is the interconnect length in units of chip footprint (see Section 3.3.1). Substituting for t_c from (3.65) and F_p from (3.66), and minimizing t_p under the assumption that $t_{RC} = 0.69R_{on}(C_o\mathcal{L}+C_L)$, i.e., a lossless interconnect is assumed to simplify the minimization problem, we obtain,

$$t_{p(min)} = 0.69 R_{on} C_L +$$

$$+ \left(\frac{27}{4}\right)^{1/3} \left(\frac{N_{cir} U R_m^2 \left(0.69 R_{on} C_o + \sqrt{\varepsilon_o \varepsilon_{eff} \mu_o}\right)^2}{q_A}\right)^{1/3} \quad (3.68)$$

A more accurate approximation for the minimum packaging delay can be obtained by assuming that the interconnect is lossy (using a form similar to that in (3.57)). Equation (3.68), however, is sufficient for illustrative purposes here. Figure 3.28 shows the minimum packaging delay as a function of the heat removal limit of the packaging computed using (3.68).

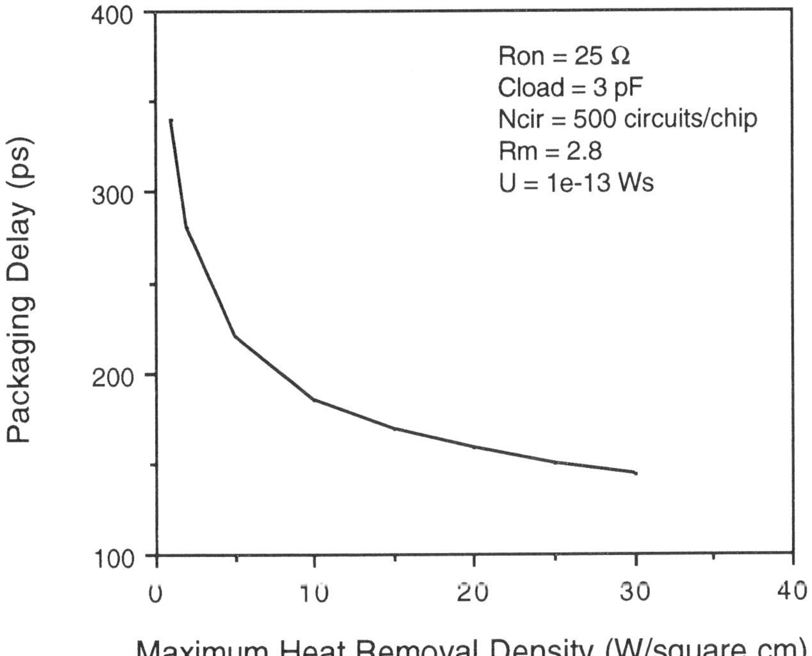

Figure 3.28. Minimum packaging delay as a function of the heat removal limit of the packaging computed using (3.68). The chip represents the series gated ECL chip previously used in Table 3.2.

Attenuation. Frequency analysis for nets can be determined from a Bode plot for the net with the driver circuit removed. While a frequency analysis of a digital signal is not generally a critical metric, knowing the attenuation associated with analog signals is of interest. The Bode plot is created by driving the net with a sinewave of variable frequency and measuring the voltage gain and phase at the receivers. The bandwidth of a transmission structure is defined as the frequency at which the output signal is attenuated by a specified amount (usually -3 dB). -3 dB represents the half power point, i.e.,

$$\text{gain (dB)} = 20\log_{10}\left(\frac{V_{out}}{V_{in}}\right) = 10\log_{10}\left(\frac{\text{Power}_{out}}{\text{Power}_{in}}\right) \tag{3.69}$$

where gain = -3 dB when $\text{Power}_{out}/\text{Power}_{in} = 0.5$.

The results from Bode plots may be used to estimate the amount by which a signal is attenuated on the net. If the driving signal is a sinewave, then the resulting attenuation may be read directly off the Bode plot for the sinewave's frequency. If the driving signal is not a sinewave, the signal can be decomposed into its Fourier components and the attenuation of each component assessed. For a trapezoidal driving signal, the following process may be followed:

1) The Fourier series representing the periodic driving signal must be determined. For the trapezoidal clock signal, assuming that the rise time and fall time of the signal are equal, the Fourier coefficients are given by [3.40],

$$V_n = 2Vd\frac{\sin(n\pi d)}{n\pi d}\,\frac{\sin(n\pi t_r/T_p)}{n\pi t_r/T_p} \tag{3.70}$$

where,

V_n = the amplitude of the nth harmonic

V = the peak-to-peak amplitude of the signal

n = the harmonic number ($n = 1$ is the fundamental)

d = the duty cycle = $(t_r + t_{on})/T_p$

t_r = the rise time (if $t_r \neq t_f$, the smaller should be used in (3.70))

t_{on} = the on time

T_p = the period = $t_r + t_{on} + t_f + t_{off}$
t_f = the fall time
t_{off} = the off time.

2) Determine the fraction of the total power in each harmonic (n = 1, 2, 3, ...) using,

$$\frac{\text{Power}_n}{\text{Total Power}} = \frac{V_n^2}{\sum_{n=1}^{\infty} V_n^2} \tag{3.71}$$

3) Determine the power gain of the nth harmonic from the Bode magnitude plot using,

$$a_n = 10^{(\text{gain}_n(\text{dB})/10)} \tag{3.72}$$

The frequency of the nth harmonic is n/T_p.

4) Solve for the attenuation of the signal on the net using,

$$\text{attenuation (dB)} = 10\log_{10}\sqrt{\Psi_1^2 + \Psi_2^2} \tag{3.73}$$

where

$$\psi_1 = \sum_{n=1}^{\infty} a_n \left(\frac{\text{Power}_n}{\text{Total Power}} \sin\phi_n \right)$$

$$\psi_2 = \sum_{n=1}^{\infty} a_n \left(\frac{\text{Power}_n}{\text{Total Power}} \cos\phi_n \right)$$

where ϕ_n is the phase of the nth harmonic from a Bode phase plot.

Noise. In high-performance systems, the management of noise is a major design issue. Noise can limit the achievable system performance by degrading edge rates, increasing delays, reducing noise margins and can cause false switching of logic gates. The noise magnitude and frequency are closely tied to the packaging and interconnect scheme used. The effects of the noise depend on logic rise and fall times, gate currents, the system size, system timing, net and bond lengths and pitches, interconnect electrical design, and power and ground distribution on the module.

In order to design a reliable system, it is necessary to understand the dc and ac noise margins. Different logic families track changing input and output levels in different ways depending on variations in temperature and supply levels. These variations can be used to determine the worst-case dc noise margins. Figure 3.29 gives a simple definition of the dc noise margin of a logic gate. If a system were to be designed around the worst case conditions then all the noise contributions summed together should not exceed the dc noise margins. This situation would, however, be far more restrictive than necessary for insuring signal integrity, and could be devastating to system performance. For example, in order to satisfy dc noise margin requirements, one could decrease load reflections by increasing the distance between the loads on a critical net which could, in turn, result in an unacceptable interconnect delay.

A better description of the noise limits for a logic gate is the ac noise tolerance or noise immunity. In particular, for each gate an input signal ac noise immunity curve, which relates not just the critical amplitude of the input noise required to falsely switch the gate, but the duration or pulse width of that noise. An example ac noise immunity curve is shown in Figure 3.30. Note that for very long pulse widths, the ac noise

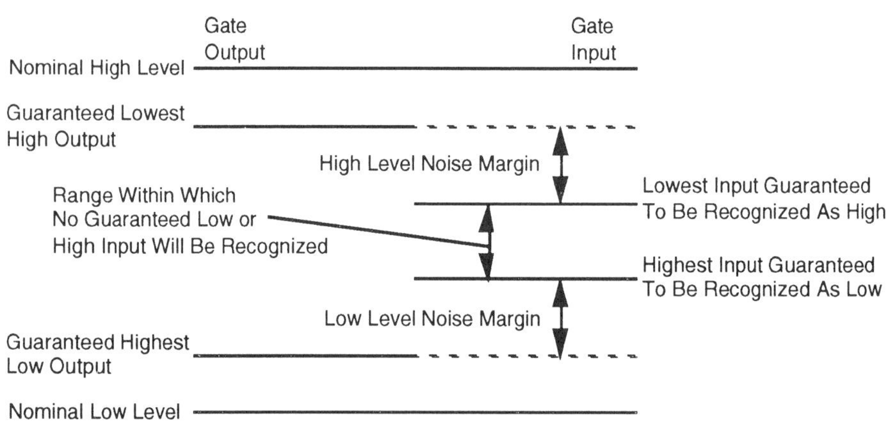

Figure 3.29. Definition of the dc noise margin of a logic gate.

immunity curve approaches the dc noise margin. The ac noise immunity is less restrictive than the dc noise margin, that is, any summation of all the noises in the system at a specific instant in time which falls below the curve in Figure 3.30 indicates that the gate will not switch falsely. The ac noise immunity relation is generated for a given logic gate by fixing the pulse width of an input signal and increasing its amplitude until the dc noise margin is exceeded by the output. Examples of the generation of ac noise immunity curves by simulation and experimentation for ECL logic are discussed in [3.41] and [3.42].

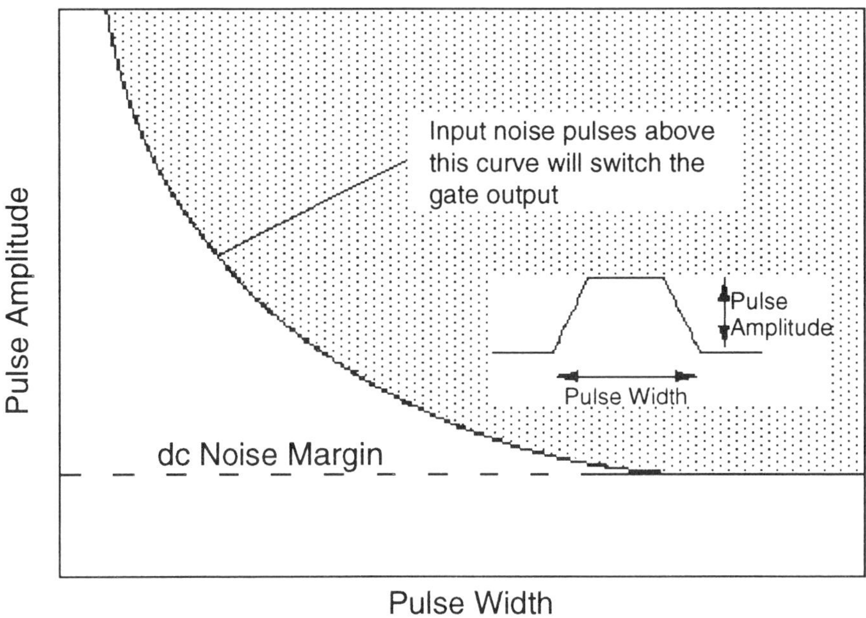

Figure 3.30. Example ac noise immunity or noise tolerance curve for a logic gate. An input noise pulse in the shaded area will falsely switch the gate. At large pulse widths the curve approaches the dc noise margin.

To set up noise management design rules for a multichip system, each noise contributing process can be allocated an acceptable noise amplitude such that the sum of all the noise amplitudes is within the ac noise tolerance. An example noise budget is shown in Table 3.3. As a benchmark, the noise amplitudes in Table 3.3 can be summed and compared to the dc noise margin. As discussed above, one does not expect the peak noise amplitudes of all the items in the noise budget to occur at the same instant in time. Therefore, the usefulness of a budget like the one in Table 3.3 is limited to identifying potential problem nets. Once problem areas are identified, detailed modeling of the specific nets taking timing information into account must be performed.

Table 3.3. Noise budget for a signal path from a driver on a thin film multichip module, through a connector to a chip on a printed circuit board. Note, in this case a parallel termination was placed on the net on the printed circuit board. If the net was either not terminated or series terminated, there would be no dc IR drop.

Noise Source	Budget (mV)	Budget (% of 5V bias level)
Delta-I Noise	150	3.0
Reflections	200	4.0
Crosstalk (signal line)	75	1.5
IR Drop (signal line)	350	7.0
IR Drop (connector)	50	1.0
IR Drop (leads, vias, etc.)	10	0.2
Internal IC Noise	50	1.0
Thermal Variations	10	0.2
Worst Case Total	895	17.9
DC Noise Margin	750	15.0

Three of the noise contributions on Table 3.3 are usually of greatest concern to module designers: crosstalk, reflections, and ΔI-noise (simultaneous switching noise). Details of the crosstalk and switching noise contributions are discussed in the following paragraphs.

Crosstalk. Crosstalk, sometimes called coupled noise, is a result of mutual inductance and capacitance between neighboring transmission lines, i.e., electric and magnetic fields associated with lines that are close to each other interact, and the lines communicate information to each other. Crosstalk increases as the distance between lines (pitch) decreases, the distance between the lines and their circuit return path (ground plane for example) increases, and the length of their close proximity (coupled length) increases.

Crosstalk is virtually impossible to realistically estimate at the conceptual design level. Because crosstalk depends on the proximity of specific signal lines, the distance that they couple, and the timing of signals on those lines, there is no way of obtaining accurate crosstalk estimates before routing a design. There is, however, a crosstalk figure of merit which can be computed and used to gain some limited insight on the comparison of technologies, i.e., saturated near end crosstalk. Under the specific termination conditions shown in Figure 3.31, when a line becomes long enough to make the propagation delay greater than half the rise time of the signal, the near end crosstalk reaches its maximum amplitude (saturates).

Under saturated conditions the line length, signal rise time, and dielectric constant of the embedding material do not affect the magnitude of the near end crosstalk, i.e., near end crosstalk is solely a function of the signal line geometry and the signal swing. The saturated crosstalk on a quiet line due to one excited line is given by Catt [3.43],

$$\text{Saturated Crosstalk (dB)} = 20\log_{10}\left|\frac{Z_{oe}-Z_{oo}}{Z_{oe}+Z_{oo}}\right| \tag{3.74}$$

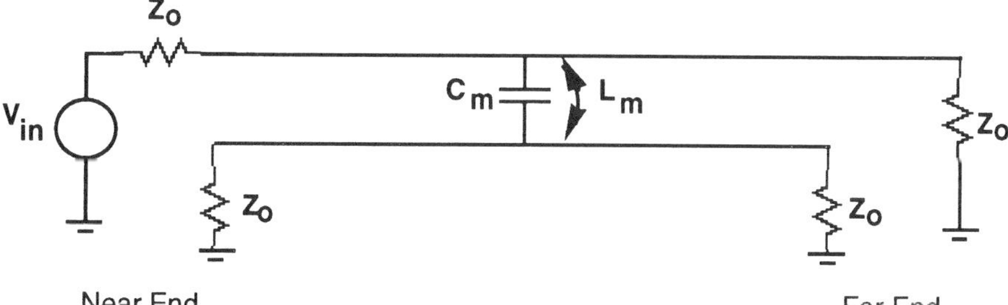

Figure 3.31. Coupled transmission lines. Both lines have a characteristic impedance of Z_O.

where Z_{oe} and Z_{oo} are the even and odd mode impedances of the coupled transmission lines given by,

$$Z_{oe} = \sqrt{\frac{L_o + L_m}{C_o - C_m}} \quad \text{and} \quad Z_{oe} = \sqrt{\frac{L_o - L_m}{C_o + C_m}}$$

where L_o and L_m are the self and mutual inductance per unit length, and C_o and C_m are the self and mutual capacitance per unit length of the coupled lines. Figure 3.32 shows the saturated crosstalk between two lines for three different 50 Ω stripline designs as predicted by (3.74). One simple way to estimate the near end crosstalk for a net that is not long enough for crosstalk to saturate is to multiply the saturated crosstalk obtained from Figure 3.32 by the fraction of the saturated length that the net represents. Following this reasoning suggests that, for non-saturated net lengths, there is an even larger difference between the potential crosstalks in printed circuit board and thin film than shown in Figure 3.32. This is because the net lengths are shorter in thin film MCMs.

The ratio of the number of excited to quiet lines also affects crosstalk and saturated crosstalk. If the number of identical excited lines next to the quiet line considered above were doubled to two, the crosstalks from each excited line would add by superposition and the total crosstalk on the quiet line would approximately double.

Crosstalk between signal interconnects is not the only source of crosstalk in a signal path. Crosstalk between bonding leads may also be a factor. The important factors in determining crosstalk on bonding leads are the mutual inductance and mutual capacitance between adjoining leads (the same as for lines). These factors are primarily dependent on the length and pitch of the leads and, to second order, on the bond geometry. For wirebonds and TAB leads of equal length and pitch, the crosstalk contribution associated with the bonding leads is nearly the same. Crosstalk on flip chip bonds will be much less than with TAB or wirebonding due to the short length of the bond. Crosstalk predictions for TAB and wirebond connections are given by [3.44].

Accurate crosstalk magnitudes can be determined by simulation of coupled-line systems. Crosstalk between lines which cross at right angles and between lines and vias or holes may need to be considered in some systems as well.

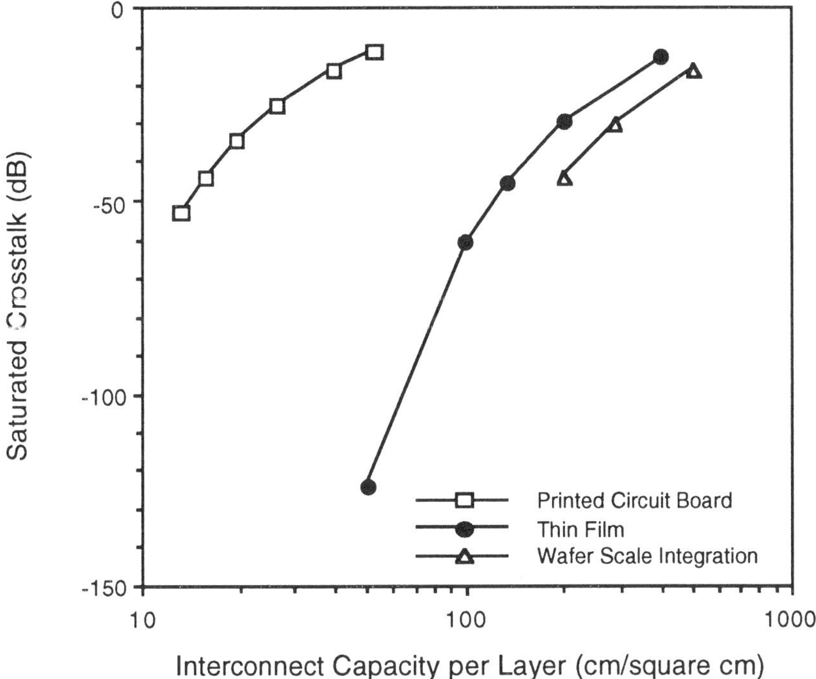

Figure 3.32. Saturated crosstalk as a function of interconnect capacity per layer for 50 Ω coupled stripline configurations. The printed circuit board had a conductor thickness = 0.7 mils, conductor width = 5 mils, ground separation (between ground planes) = 15 mils, and a dielectric constant = 4.5. The thin film interconnect had a conductor thickness = 0.2 mils, conductor width = 0.6 mils, ground plane separation = 1.7 mils, and a dielectric constant = 3.4. The wafer scale integration example had a conductor thickness = 2 μm, conductor width = 10 μm, ground plane separation = 30 μm, and a dielectric constant = 3.9.

Simultaneous Switching Noise. A potentially serious problem faced by high performance systems is switching or delta-I noise. This type of noise originates when drivers switch current drawn from a power distribution system.

Switching noise is defined by the expression for the voltage drop across an inductor,

$$V_{noise} = nL_{eff}\frac{di}{dt} \tag{3.75}$$

where,

V_{noise} = switching noise magnitude
n = number of simultaneously switching drivers
L_{eff} = effective inductance of the path through which di/dt flows
di/dt = current slew rate of one driver.

Like crosstalk, switching noise is a system design dependent quantity. The contributions to switching noise at the module level come from the inductance of bonding leads, single chip packages, on-module power/ground delivery system (i.e. planes, buses), and power/ground delivery to the module.

Bonds usually represent the most immediate source of inductance to a switching gate. It is important to decrease the inductance of the bonding leads as much as possible. The two ways of reducing this inductance are to make the bonds short and to increase the number of power and ground leads (decreasing the signal to ground ratio) placing the leads which comprise the circuit return path (usually ground leads) as close to the leads carrying the switching current as possible, thus increasing the mutual inductance component associated with the bonds. The effects of signal/ground/power lead distribution and bond pitch are discussed in detail by Rainal [3.45]. As an example, the effective inductance associated with a ground lead can be estimated by considering the expression for the effective inductance of a ground lead which can be written as, [3.46],

$$L_{eff} = \frac{L_{self}I_{ground} - (\sum L_{mutual}I_{signal})}{I_{ground}} \tag{3.76}$$

where L_{self} is the self inductance of a ground lead, $\sum L_{mutual}$ is the mutual inductance of the ground lead due to I_{signal} flowing in the surrounding signal lines, and I_{ground} is the current flowing in the ground lead. Note that the ratio of I_{signal} to I_{ground} is the signal to ground lead ratio. Figure 3.33 shows the effective inductance plotted as a function of lead length for various bonding approaches.

The potential for less switching noise when flip chip attach approaches are used is obvious from Figure 3.33. Increasing the signal to ground ratio has the effect of decreasing the mutual inductance and thereby increasing L_{eff} and V_{noise}, and the overall result is to shift the curves in Figure 3.33 upward. Increasing the pitch of the leads decreases the mutual inductances and will also shift the curves in Figure 3.33 upwards.

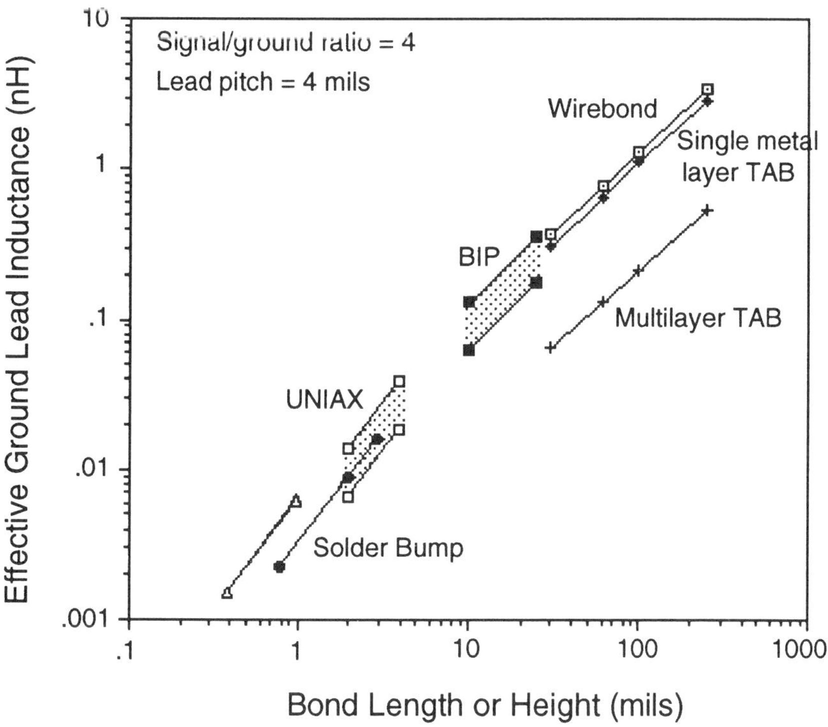

Figure 3.33. Effective inductance as a function of bond length for various bonding technologies. The signal to ground ratio and lead pitch was held constant in all cases. The following specific dimensions were assumed: Wirebond - diameter = 1 mil, Single metal layer TAB - width = 2 mil, thickness = 1 mil, Multilayer TAB - separation between traces and ground ring = 2 mils, BIP (Bonded Interconnect Pin) [3.47] - diameter = 17 to 75 μm, UNIAX [Raychem Limited] - diameter = 25-75 μm, Solder Bump - diameter = 75 μm.

The effective inductance for lead paths through single chip packages range from 3.5 nH for a C4 flip chip bonded die in a metalized ceramic PGA single chip package (100 mil pin pitch) to over 20 nH for a chip wirebonded into a plastic leaded chip carrier (50 mil lead pitch), [3.48] ([3.48] and [3.49] give inductance estimates for numerous other package configurations). The reader should be cautioned that for a given package/lead configuration, a range of inductances can be observed depending on signal, power, and ground assignments for the particular application.

To formulate the overall L_{eff} for a module design, estimates of the inductance associated with the bonding technology, chip packages, and interconnect power distribution network are necessary. The accurate estimation of switching noise at the module level generally requires detailed timing information which may not be available at the conceptual design level. Techniques for estimating L_{eff} for module designs are outlined in [3.50] and [3.51].

The management of switching noise can be approached in a number of ways. The most direct method is to attack the problem on the chip by using complementary or balanced driver logic implementations. In principle, a pair of complementary outputs should contribute equal but opposite noise voltages to the line to which they are terminated and the opposing noise voltages should to cancel. The effectiveness of using complementary outputs for the management of switching noise in ECL systems is discussed by [3.52]. Active regulation schemes, usually taking the form of a current source, and passive bypass capacitors implemented on the chip can also be used to regulate noise. Alternatively, on-chip bypassing can be used, [3.9].

Numerous options also exist for managing switching noise using off-chip methods, the most common of which is the use of bypass capacitors. Bypass capacitors should ideally be placed such that the amount of inductance between the capacitor and the switching event is minimized. One approach is to place bypass capacitors on the die connected to the inner lead bond pads, [3.53]. Bypass capacitors may also be placed in or on a single chip package or on the interconnect or board. However, in these cases, the effective inductance of the bonding leads will not be bypassed. If bypass capacitors are placed on the interconnect, the use of low impedance interconnects has the advantage that the bypass capacitor's proximity to the switching event is not critical as long as they are located close enough that the delay between the gate and the bypass is much less than the signal rise and fall times. Thin and thick film interconnects have the added advantage that bypass capacitors can be built into the interconnect and the use of closely coupled power and ground planes can often supply much of the needed bypass capacitance.

Analytical Approaches to Modeling Switching Noise. It has been shown that while switching noise scales linearly with the number of drivers for small noise magnitudes, the simple relation in (3.75) can overestimate the switching noise by as much as 100% if the di/dt ratio is not adjusted as n increases [3.54, 3.55]. While approaches based on (3.75) can be used to estimate noise magnitudes and the amount of bypass capacitance needed to manage switching noise in some cases, accurate solutions to power distribution problems in multichip modules where n is large require either careful modification of the current slew rate in (3.75) or the use of system simulations.

A number of authors have extended (3.75) to include bypass capacitors [3.56, 3.57]. Val and Martin [3.56] have derived a relation for computing the switching noise magnitude as a function of time. Their relation is dependent on the capacitance and series parasitics (resistance and inductance) associated with a bypass placed between power and ground in a simple TTL system:

$$V_{noise}(t) = \frac{I_{max}}{t_r}\left(L + Rt + \frac{t^2}{2C}\right), \quad \text{for } t < t_r \tag{3.77}$$

where,

t_r = rise time
I_{max} = peak (total) switching current in the system
L = interconnect + bypass series inductance
R = interconnect + bypass series resistance
C = bypass capacitor value
t = time.

Hashemi *et al* [3.57] have derived a similar relation for predicting switching noise in CMOS systems with bypass capacitors. The derivation in [3.57] is slightly more detailed than the derivation leading to (3.77), treating the interconnect inductance separately from the bypass series inductance and uses an explicit form for I_{max} dependent on the characteristics of the CMOS drivers:

$$I_{max} = \frac{n}{t_r}\left(\frac{\mu C_{ox} w}{2l}(V_{dd} - V_t)^2\right)t = \xi t \qquad (3.78)$$

where the following quantities are defined for the NMOS transistor,

μ = surface mobility
C_{ox} = capacitance per unit area of the gate oxide
l = channel length
w = channel width
V_{dd} = gate bias
V_t = threshold voltage.

From the simple circuit shown in Figure 3.34 (neglecting R_o) and (3.78) the time dependent noise voltage becomes,

$$V_{noise}(t) = L_o \xi\left(1 - \frac{L_o}{L_o + L}e^{-\alpha t}\left(\cos\beta t - \frac{\alpha}{\beta}\sin\beta t\right)\right) \qquad (3.79)$$

where,

$$\alpha = \frac{R}{2(L + L_o)}$$

$$\beta = \sqrt{\frac{1}{(L + L_o)C} - \alpha^2}$$

The approximation for I_{max} used in both (3.77) and (3.78) assumes that the current variation during switching has a triangle shape, i.e., in the interval of $0<t<t_r$ the derivative of the current with respect to time is a constant. The approximations for I_{max} also assume direct proportionality with the number of switching gates resulting in both derivations overpredicting the noise magnitude in most cases of interest (see Figure 3.35).

A noise estimation similar to (3.79) has been developed by Canright [3.58]. The solution is also based on the circuit shown in Figure 3.34

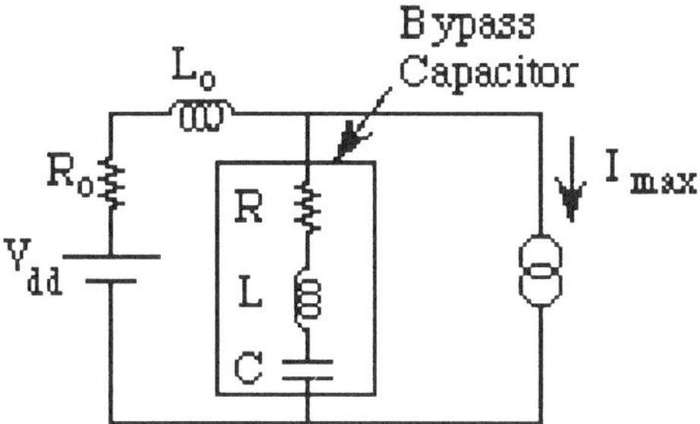

Figure 3.34. Simplified CMOS system equivalent circuit. I_{max} is the total current demand from n CMOS drivers given by (3.78). R, L, and C represent the bypass capacitor, L_o is the effective inductance of the interconnect and leads, V_{dd} is the system bias, and R_o is the parasitic resistance of the interconnect, leads, and power supply.

where R and L are considered negligible (but R_o is included) and I_{max} is assumed to have an exponential form,

$$I_{max}(t) = \frac{C_{ox}V_{dd}}{t_r}e^{-t/t_r} \tag{3.80}$$

The resulting noise and its derivation appear in [3.58].

A simple noise prediction equation for CMOS circuits which includes the effects of feedback in the expression for I_{max} and therefore results in noise saturation has been developed by [3.54] using,

$$I_{max} = \frac{n}{t_r}\left(\frac{\mu C_{ox}w}{2l}(V_{dd} - V_t - V_{noise}(t))^2\right)t \tag{3.81}$$

The complete expression for the noise at $t=t_r$ becomes,

$$V_{noise} = \frac{b - \sqrt{b^2 - 4(V_{dd} - V_t)^2}}{2} \tag{3.82}$$

where,

$$b = 2\frac{t_r}{L_o}\left((V_{dd} - V_t)\frac{L_o}{t_r} + \frac{1}{n\mu C_{ox}w}\right)$$

Equation (3.82) is more accurate than (3.75) and (3.77)-(3.79) however lacks the detail to be useful for a large number of real designs, i.e., (3.82) is hard to use with bypass capacitors and does not explicitly separate interconnect, bonding, and other parasitics.

Following the derivation of [3.57] and including feedback in the prediction of the output driver current as done in (3.81) an alternative analytical noise prediction expression described by the simultaneous solution of (3.83a)-(3.83c) can be developed,

$$L\frac{d^2 i_2}{dt^2} + R\frac{di_2}{dt} = L_o\frac{d^2 i_1}{dt^2} + R_o\frac{di_1}{dt} - \frac{i_2}{C} \tag{3.83a}$$

$$i_1 + i_2 = \frac{n}{t_r}\left(\frac{\mu C_{ox} w}{2l}\left(V_{dd} - V_t - L_o\frac{di_1}{dt}\right)^2\right)t \tag{3.83b}$$

$$V_{noise}(t) = L_o\frac{di_1}{dt} \tag{3.83c}$$

The predictions of (3.79), (3.82), and (3.83) for a nine layer medium film copper/polyimide multichip module containing three chips are shown in Figure 3.35. More details on the construction, measurement, and modeling of this module are provided in [3.52]. It should be noted that while the predictions from (3.82) and (3.83) match the actual noise results better than the other formulations, the peak noise magnitude is still significantly overpredicted in all cases. It should also be noted for large n in Figure 3.35 the data points were obtained from detailed simulation of the module and experimental verification was not available. Our past experience has indicated that even detailed module models often tend to overpredict rather than under predict switching noise (due to the assumption of exact simultaneity of switching in the simulations) and the actual switching noise may be a few percent less than that shown in Figure 3.35.

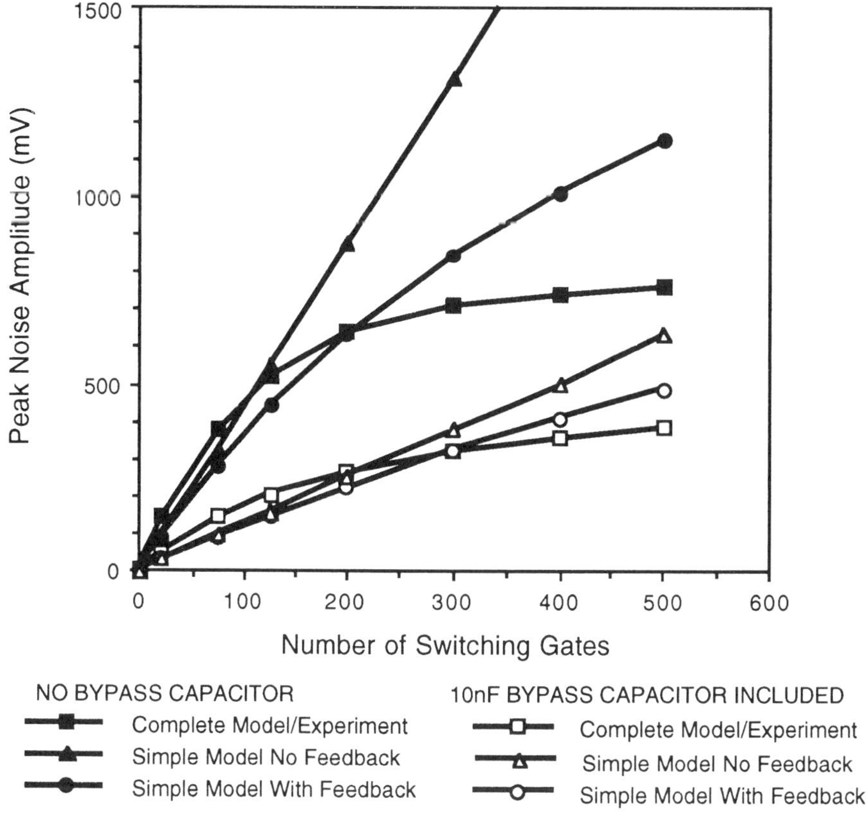

Figure 3.35. The performance of various simple switching noise models [3.52]. The "Complete Model/Experiment" results are a combination of experimental measurements and a detailed circuit model calibrated to the measurements. The simple model with no bypass capacitor and no feedback can be obtained from (3.75) or (3.79) modified to include R_0. The simple model with no bypass capacitor and feedback included can be obtained from (3.82) modified to include R_0 or (3.83). The simple model with the bypass capacitor and no feedback comes from (3.79) modified to include R_0, and the simple model with the bypass capacitor and feedback included comes from (3.83).

DC Drop. A potential problem in high power modules is unwanted dc drops in the power distribution system. DC drops can add to crosstalk and switching noise to reduce receiver noise margins. The dc drop problem is sensitive to the number and characteristics of off-module power and ground connections as well as the characteristics of the power distribution system on the module. A simple dc model can be used to estimate the dc drop as a function of position on the reference planes.

The simple model in Figure 3.36 is used as the basis for the dc drop analysis. The first step in performing the analysis is to define a uniform grid which covers the whole module. Next a resistor mesh which models the power and ground planes is constructed. The value of each resistor in the mesh in the x and y directions in Figure 3.36 is given by,

$$R_x = \frac{w_x \rho \, grid_y}{2 t w_y grid_x f_m} \tag{3.84a}$$

$$R_y = \frac{w_y \rho \, grid_x}{2 t w_x grid_y f_m} \tag{3.84b}$$

where,

w_x and w_y = the width of the module in the x and y directions

ρ = the resistivity of the reference plane metal

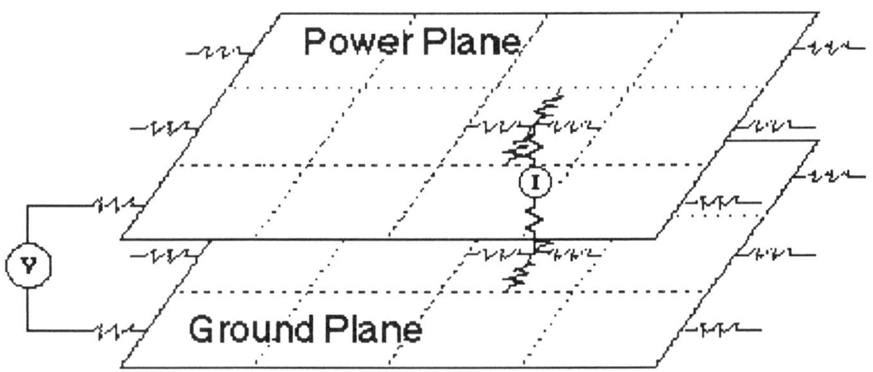

Figure 3.36. Equivalent circuit model constructed for the dc drop analysis.

$grid_x$ and $grid_y$ = the number of cells in the x and y direction
t = the thickness of the reference plane metal
f_m = the fraction of metal in the reference planes (1.0 = solid plane, 0.5 = 50% metal mesh)

Resistors are attached to the edges of the module which represent the off-module connectors. The resistance of a connector trace is determined using,

$$R_{\text{connector trace}} = \frac{length_{ct}\rho_{ct}}{area_{ct}} \tag{3.85}$$

where,

$length_{ct}$ = length of a connector trace
ρ_{ct} = the resistivity of the connector trace metal
$area_{ct} = w_{ct}\, t_{ct}$, cross-sectional area of a connector trace
w_{ct} = width of a connector trace
t_{ct} = thickness of a connector trace.

The values of the resistors attached to the edge of the module model are given by,

$$R_{\text{per cell}} = R_{\text{connector trace}} \frac{pitch_{ct}\, grid_{x\text{ or }y}}{w_{x\text{ or }y}} \tag{3.86}$$

where

$pitch_{ct}$ = pitch of the connector traces.

After the resistive mesh is completed, the next step is to evaluate the positions of the components in the module. For the ith cell the set of components in the cell (either completely or partially) is evaluated to determine the amount of power dissipation in the ith cell. The current source attached to the ith cell is then determined using,

$$I_i = \frac{P_{c_i}}{V_{supply_i}} \tag{3.87}$$

where

P_{c_i} = the power dissipation in the ith cell

V_{supply_i} = the supply voltage for chips in the ith cell.

The estimation described above assumes that a component's power dissipation is uniform over the entire area of the component.

3.3.5 Dependability Analysis

Dependability refers to the ability of a system to accomplish the tasks which are expected of it. Measures of dependability are aimed at quantifying the quality of service which is provided to the user with respect to error manifestation and processing. Specific metrics used to measure dependability include: reliability, maintainability, and availability [3.59]. This section concentrates primarily on reliability.

Reliability is the statistical art of determining the length of time over which a device or system will function correctly. Reliability of electronic systems depends on both the hardware, such as integrated circuits, single chip packages, bonding techniques, interconnects, cooling schemes, and power supplies; and the software, such as operating systems and application programs. The cost of repairing failures has been rising almost as quickly as integrated circuit costs have dropped, so reliability may well be the most important design goal for many systems - including inexpensive consumer electronics applications.

Unlike mechanical components, modern electrical components rarely wear out. Therefore, failures can be predicted only on a statistical basis. The faults which electronic systems encounter can be categorized as:

1) Physical Faults. Physical faults are caused by adverse physical phenomenon, and can be either internal to the system, resulting from a physical-chemical disorder (electromigration, oxide breakdown, etc.) or external to the system, induced by environmental perturbations (thermal shock, vibration, etc.).

2) Man-Made Faults. Man-made faults are either design defects introduced into the system during initial design or later modification, or interaction and maintenance faults resulting from improper use of the system or improper maintenance.

It should be noted that faults of both types can combine to produce an error.

Reliability is a system characteristic which must be built into the design, it can not be added after product development. Therefore, aggressive design for reliability can result in designs which have orders of magnitude improvement in system reliability. Design for reliability can be accomplished using several techniques: fault avoidance, fault detection, fault masking, and redundancy. Redundancy is discussed in more detail at the end of this section. Design for reliability is treated in more detail in [3.60] and its associated references.

Dependability Metrics. The reliability function R(t) of a component or system represents the conditional probability that it will operate correctly at time t, given that it is operational at time zero. In a system without repair (after leaving the factory), the most commonly used reliability function is the exponential function,

$$R_c(t) = e^{-\lambda_c t} \tag{3.88}$$

where λ_c is the failure rate expressed in failures per specified time period, t represents the time period, and $R_c(t)$ is the probability of correct operation at time t.

To predict the reliability of a system of components, we adopt a bottom-up approach, i.e., the system reliability is an integration of the reliabilities of the lower-level subsystems or components which make up the system. Non-redundant systems must have all their components operational for the system to function properly, this type of system's reliability is given by,

$$R_s(t) = \prod_{i=1}^{n} R_{ci}(t) = e^{-\lambda_s t} \tag{3.89}$$

where $R_s(t)$ is the probability that the system will be operational at time t and $R_{ci}(t)$ represents the reliability of the ith component (out of a total of n). If the reliability function is exponential, the system failure rate is given by,

$$\lambda_s = \sum_{i=1}^{n} \lambda_{ci} \tag{3.90}$$

Mean Time To Failure (MTTF). The MTTF is the average period of time for which a system will operate before failing relative to time zero. The MTTF for a system is given by,

$$\text{MTTF} = \frac{1}{\lambda_s} \tag{3.91}$$

Mean Time Between Failures (MTBF). The MTBF is meaningful for systems where repair times are also considered. If the failure rate is constant over time and the mean time to repair a failure (MTTR) is negligible, the MTBF is the same as the mean time to the first failure (MTTF). For a non redundant system the MTBF is given exactly by,

$$\text{MTBF} = \text{MTTF} + \text{MTTR} \tag{3.92}$$

MTTR becomes critical in determining the MTBF of redundant systems. The formulation for determining MTTR is given in (3.96).

Mission Time. Mission time (t_m) is the time at the end of which the probability that the system is still operational is a constrained value, r. Mission time is a critical metric for describing systems where continued reliable operation in the absence of repair is vital (such as a satellite). For non-redundant systems with constant failure rates over time, the mission time is given by,

$$t_m = \frac{-\ln(r)}{\lambda_s} \tag{3.93}$$

Availability. Availability is the percentage of time a system is operational. Many classes of systems can tolerate brief downtimes, but must be operational the majority of the time (for example an automated bank teller). Availability is computed using,

$$A_i = \frac{\text{MTTF}}{\text{MTTF} + \text{MCT}}, \quad \text{inherent availability} \tag{3.94a}$$

$$A_a = \frac{\text{MTBM}}{\text{MTBM} + \text{MTTR}}, \quad \text{actual availability} \tag{3.94b}$$

where MTBM is mean time between maintenance and MCT is the

mean corrective time. The MTBM is given by,

$$\text{MTBM} = \frac{1}{\dfrac{1}{\text{MTTF}} + \sum_{j=1}^{m} pm_j} \tag{3.95}$$

where pm_j is the frequency of the jth preventative maintenance action and m is the total number of preventative maintenance actions performed. The mean time between maintenance (MTBM) for a system with no preventative maintenance is equal to the MTTF. The mean time to repair is given by

$$\text{MTTR} = \text{MTBM}\left(\frac{\text{MCT}}{\text{MTTF}} + \text{MPT}\sum_{j=1}^{m} pm_j\right) \tag{3.96}$$

where MPT is the mean preventative maintenance time. MCT and MPT are given by

$$\text{MCT} = \text{MTTF}\sum_{i=1}^{n} \lambda_{ci}\text{MCT}_i \tag{3.97}$$

$$\text{MPT} = \frac{\sum_{j=1}^{m} pm_j \text{MPT}_j}{\sum_{j=1}^{m} pm_j} \tag{3.98}$$

where MCT_i is the mean corrective time for the ith component and MPT_j is the mean preventative maintenance time for the jth action.

Error Coverage. Error coverage is the fraction of errors which a system is capable of detecting and reporting as they occur. Prompt detection of errors can lead to shorter MTTR and higher availability. Error coverage is determined by a combination of system software and hardware design. The computation of error coverage requires knowledge of failure modes and their associated probabilities of occurrence [3.61].

Failure Rate Analysis Methods. The analysis of reliability concentrates on the determination of failure rates (λ). Three distinctly different failure rate regions are generally displayed by electronic systems (Figure 3.37): 1) early failures (infant mortality) due to defect escapes from the manufacturing process, 2) constant intrinsic failure rate period characterized by random failures, and 3) wear out due to failure mechanisms like conductor electromigration. The functions presented earlier in this section, (3.88)-(3.90) only model the operational life period, not the infant mortality or wear out periods.

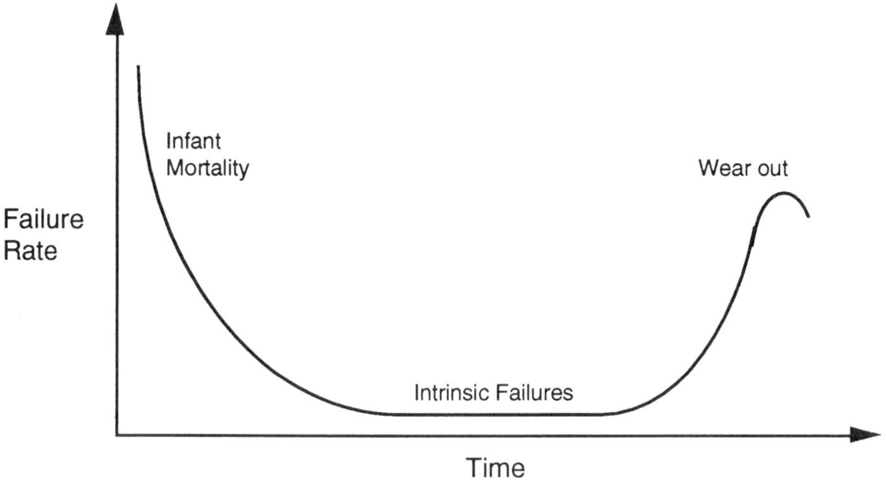

Figure 3.37. Failure rate versus time, "bathtub" curve.

There are several methods for determining failure rates of individual components and packaging elements. Models developed to predict the effect of temperature on chemical reaction rates have been widely applied to electronic components. The two most prevalent models are the Arrhenius and Eyring models. These models use exponential failure distributions (and therefore constant failure rates), and generally make the assumption that the dominant component failure mechanisms depend on the steady state temperature. This assumption is used to derive activation energies or thermal acceleration factors for the component based on a weighted averaging methodology. The failure rates in these cases are characterized as an exponential function of the steady state temperature:

Arrhenius Form:

$$\frac{1}{\lambda} = K_1 e^{E_a/kT} \tag{3.99}$$

Eyring Form:

$$\frac{1}{\lambda} = K_2 e^{E_a/kT} e^{\beta/rh} \tag{3.100}$$

where,

rh = relative humidity
E_a = activation energy (electronvolts)
k = Boltzmann constant
T = temperature of the component
K_1, K_2, and b = constants.

The exponential functions in (3.99) and (3.100) model the operational life period only, not the infant mortality or wear out periods.

The best known reliability prediction methods are published in MIL-HDBK-217 [3.62]. MIL-HDBK-217 failure rate predictions embed Arrhenius models and take the form

$$\lambda = \lambda_b \prod_i \pi_i \tag{3.101}$$

where λ_b is the component's base failure rate and π_i are acceleration factors which account for variations in quality, temperature, voltage, and environment. MIL-HDBK-217 models have been shown to be a conservative estimator of absolute failure rate (i.e., they predict higher failure rates than those actually observed), but, if used correctly, they can provide predictions of relative failure rates among different systems. A discussion of the shortcomings of MIL-HDBK-217 may be found in [3.63].

Exponential failure distributions do not accurately model all failure mechanisms under all temperature variations. Some failure mechanisms require other failure distributions in order to be accurately described, i.e., normal, lognormal, and Weibull distributions have been used. A comprehensive discussion of the techniques for statistical treatment of reliability data is available in [3.64].

MIL-HDBK-217 reliability prediction methods do not account for temperature dependencies other than steady state temperature. Some failure mechanisms have inverse temperature dependencies, temperature cycle or gradient dependencies, or time dependent temperature change dependencies. In addition, the use of a single activation energy is in general inappropriate for accurate failure rate predictions.

An alternative to the statistically-based failure rate predictions in MIL-HDBK-217 are physically-based models. Physics-of-failure methods [3.65] model actual failure mechanisms to determine failure rates and do not depend on the existence of field data to provide predictions. Therefore, new materials and technologies are not penalized because of a lack of field failure data. Physics-of-failure is an up-front approach to reliability which utilizes the knowledge of stresses, materials, and structure to identify potential failure mechanisms. Where a stress refers to the impact of environmental and operating conditions, such as an applied force or an electric field. A failure mechanism refers to the physical process(es) that bring about failure, such as electromigration, corrosion, or fatigue.

Redundancy (Design for Reliability). Redundancy can be included in a system to handle errors and improve the overall system reliability. For example, consider a system where two processors are used to provide a backup in case of failure. If the MTTF for each processor alone is 500 hours and the MTTR of a processor is 1 hour, the MTBF of the dual processor system is:

$$\text{MTBF} = \frac{(501 \text{ hours})^2}{1 \text{ hour}} = 2.5 \times 10^5 \text{ hours}$$

The improvement in reliability due to redundancy can be tremendous.

Redundancy techniques include fault detection, fault masking, and dynamic redundancy. Fault detection techniques report the occurrence of a failure and prevent the propagation of erroneous data. Depending on the needs of the system, these techniques may be used to isolate failures to a small portion of the system hardware to facilitate the rapid location and replacement of failed components. Fault masking is accomplished by adding additional hardware to hide the presence of a fault. A simple example is triplication of a subsystem and voting so that a fault in one copy is outvoted by the other two. Dynamic redundancy includes spare hardware which can replace failed components or subsystems.

In redundant systems, failures only occur when all the redundancy is exhausted. In the case of systems which include redundancy, the system reliability is given by

$$R_s(t) = \prod_{i=1}^{M} R_{sub_i}(t) \tag{3.102}$$

where the product is over M subsystems which may have redundancy built in.

For a subsystem in which K out of N identical, non-repairable versions of the subsystem must be operational at a time, the reliability and MTTF are given by

$$R_{sub}(t) = \sum_{i=K}^{N} \binom{N}{i} e^{-\lambda_{sub} ti} \left(1 - e^{-\lambda_{sub} t}\right)^{N-i} \tag{3.103a}$$

where,

$\binom{N}{i}$ = number of possible combinations of N objects taken i at a time

λ_{sub} = failure rate of one redundant subsystem.

The MTTF is found by integrating the $R_{sub}(t)$ over all time,

$$MTTF = \int_0^{\infty} R_{sub}(t)\, dt = \frac{1}{\lambda_{sub}} \left(\sum_{i=K}^{N} \frac{1}{i} \right) \tag{3.103b}$$

For a redundant subsystem in which one out of N not necessarily identical, non-repairable versions of the subsystem must be operational at a time, the reliability and MTTF are given by

$$R_{sub}(t) = 1 - \prod_{i=1}^{N} \left(1 - e^{-\lambda_{subi} t}\right) \tag{3.104a}$$

$$\mathrm{MTTF} = \sum_{i=1}^{N} \sum_{j=1}^{C(N,i)} \frac{1}{\sum_{k=1}^{i} \lambda_m} \qquad (3.104b)$$

where,

t = time

$R_{sub}(t)$ = probability that the subsystem is operational at time t

λ_{subi} = failure rate of the ith redundant subsystem

N = total number of non-identical, non-repairable versions of the subsystem (of which one must work)

λ_m = the failure rate of the kth component in the jth combination set

$C(N,i) = \binom{N}{i} = \frac{N!}{i!\,(N-i)!}$, combination of N objects taken i at a time.

3.3.6 Cost Analysis

The potential performance advantages of multichip modules over conventional printed circuit board packaging have been demonstrated many times. While multichip modules do not perform significantly better for every application, they almost never underperform conventional approaches which use single chip packages. If the performance of multichip modules is potentially superior, why are so few applications committed to volume production using multichip modules? The reason has little to do with size, electrical, or thermal performance, but is a result of manufacturing issues including cost, lack of infrastructure, and the reluctance of manufacturers to change to new technologies. Manufacturing issues at the module level include: cost, testability, time-to-build, repair and rework, manufacturability, learning curve, and the establishment of infrastructure.

Cost is the most important performance metric in many packaging and interconnect systems. The relative costs of bonding techniques and interconnect technologies are available from numerous sources and these comparisons provide important tradeoff information. However, it is the completed system costs including rework and testing which are the ultimate determining factor in cost analysis.

To completely assess the cost of a potential system the following elements must be considered:

- Non-recurring design and development costs
- Learning curves
- Capital costs
- Tooling costs
- Recurring fabrication and assembly costs of components and packaging
- Repair and rework costs
- Testing costs
- Support and maintenance costs

It is outside the scope of this book (and practical conceptual design) to present a detailed dissertation on how each process step in a multichip system fabrication and assembly process is costed. The focus of this treatment is on a high level approach to comparing the costs of multichip systems. The following subsections treat the estimation of die costs, interconnect fabrications costs, and overall module costs including test and rework.

Bare Die Costs. Once die sizes are determined, the cost of a die can be estimated. It is necessary to determine the values of several parameters in order to determine the die cost, including the processed wafer cost, the defect density on the wafer, and the number of die which can be fabricated on a single wafer. We usually fix the processed wafer cost at the conceptual design level for systems. The fixed value depends on the type of die being fabricated, the wafer size, and several other parameters. Processed wafer costs range from \$500 to \$800 for a 6 inch diameter wafer (for a submicron CMOS process).

The next element to determine is the number of die which can be fabricated on a wafer (number up),

$$\text{Number Up} = \frac{\pi (d_w - b)^2}{4 (w_{die} + s)(l_{die} + s)}$$

$$- \frac{\pi (d_w - b)}{\sqrt{2 (w_{die} + s)(l_{die} + s)}} - \text{number of test die per wafer} \qquad (3.105)$$

where,

d_w = the wafer diameter
b = unusable wafer border width
w_{die} = die width
l_{die} = die length
s = minimum spacing between die.

The first term in (3.105) is the ratio of the usable wafer area to the die footprint, the second term takes into account that the wafer is round and the die are rectangular, and the third term subtracts the area of test die that may be strategically placed on the wafer.

Knowing the wafer cost and number up allows the estimation of the un-yielded/un-tested/un-burned-in cost per die. The last piece of information needed about the die is its yield on the wafer. The yield of die on a wafer can be computed from the defect density using a selected probability distribution. The most appropriate probability distribution to use depends on the defect density and die size. Figure 3.38 shows the regions of applicability of three applicable probability distributions.

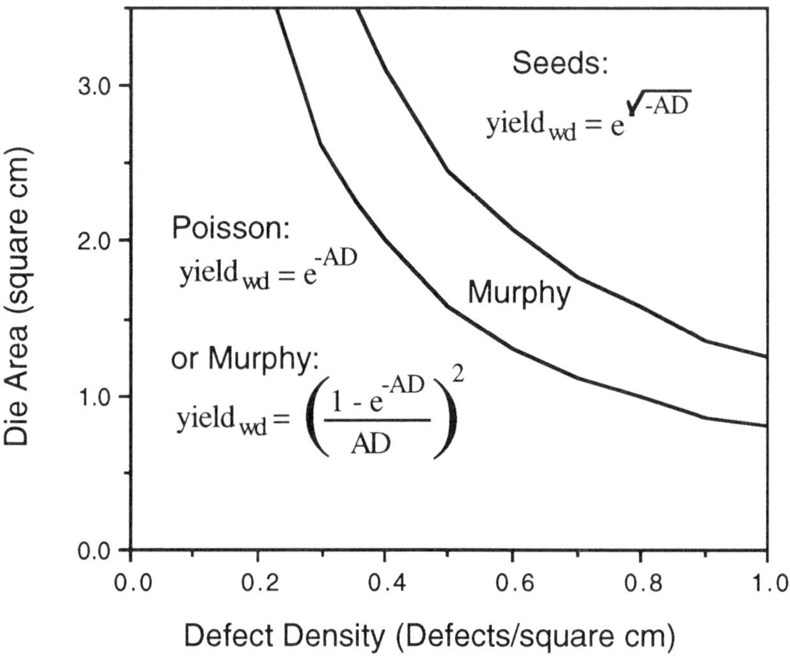

Figure 3.38. Applicability of probability distributions for estimating the yield of die on a wafer. $yield_{wd}$ is the resulting yield of the die on the wafer, A is the die area, and D is the average defect density on the wafer. The distributions included on the figure are discussed in more detail in [3.66] and [3.67].

To compute the yield and cost of the die at the beginning of the assembly process, a methodology like the one shown in Figure 3.39 can be followed. Depending on the bonding technology used, the wafer may, or may not have to be bumped (bumping is necessary for metallurgical flip chip and TAB). Die level burn-in and test are optional activities which may or may not be practical depending on the application (see the Known Good Die discussion at the end of this section).

The wafer yield of the die, discussed above, represents the actual die

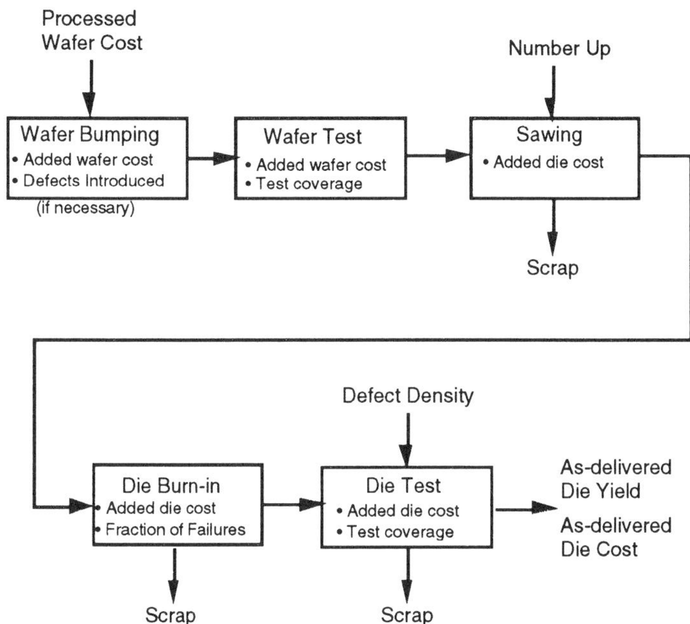

Figure 3.39. Simple die cost model.

yield. Since not all the defective die can be identified at wafer level test or die level test and burn-in, some defective die will be assembled into modules. The die yield after wafer test is computed from the test coverage (fraction of the defects identified in the test) and the actual yield of the die on the wafer,

$$\text{yield}_{\text{die}} = (\text{yield}_{\text{wd}})^{(1 - \text{test coverage})} \tag{3.106}$$

Equation (3.106) is based on the Williams and Brown model [3.68][5].

5. The yield predicted by (3.106) is higher than the yield intuitively found if the test coverage referred to the number of defective die identified in the test. Equation (3.106) is derived by accounting for the possibility that a defective die could have more than one defect, but that the identification of any defects, not necessarily all defects, is enough to scrap a die. A more accurate formulation for this yield has been derived, [3.69], however it depends on an additional parameter that accounts for clustering of defects (i.e., the average number of faults per faulty part) which is not generally known or estimatable at the conceptual design level.

After sawing, the known defective die are scrapped and the remainder are sent on to burn-in. The fraction of die which continue to burn-in and test is given by,

$$\text{pass fraction} = (\text{yield}_{wd})^{\text{test coverage}} \tag{3.107}$$

The burn-in step is characterized by a cost and a fixed fraction of die which fail during burn-in. The die test step is treated using (3.106) and (3.107) by replacing yield_{wd} with the die yield at the beginning of the step. The wafer and die cost are computed at the end of each of the steps shown in Figure 3.39 using the relation,

$$\text{cost}_{\text{cummulative}} = \frac{\text{cost}_{\text{previous step}} + \text{cost}_{\text{step}}}{\text{pass fraction}} \tag{3.108}$$

Note that all costs prior to sawing the wafer into separate die (wafer processing, wafer level testing, sawing, etc.) must be divided amongst the die which continue through the process after sawing. The impact of die yield on module costs is treated later in this section.

If die are placed into single chip packages the price of the single chip package must be estimated. The relation in Figure 3.40 can be used.

Interconnect Costs. An interconnect is defined as the module or board which connects multiple chips together. The cost of interconnects have been compared using a number of different measures, few of which, in themselves, allow an "apples-to-apples" comparison of technologies (i.e., no simple application dependent comparisons are available).

The most common method is to compute the cost per unit area of interconnect. This measure can be misleading because it does not take into account that an application packaged with a high wiring density technology (thin film for example) may require a smaller interconnect area than if it were packaged using a low density technology. Cost per area can be used to describe an interconnect only if it is accompanied with data on the resulting module size. Cost per area can also be deceiving for technologies which require the drilling of holes since their cost may be highly dependent on the number of holes which must be drilled. Figure 3.41 shows this type of interconnect cost relation. Versions of Figure 3.41 which contain more design detail (i.e., layer counts, one sided versus two sided, etc.) have been generated [3.70]. These relations

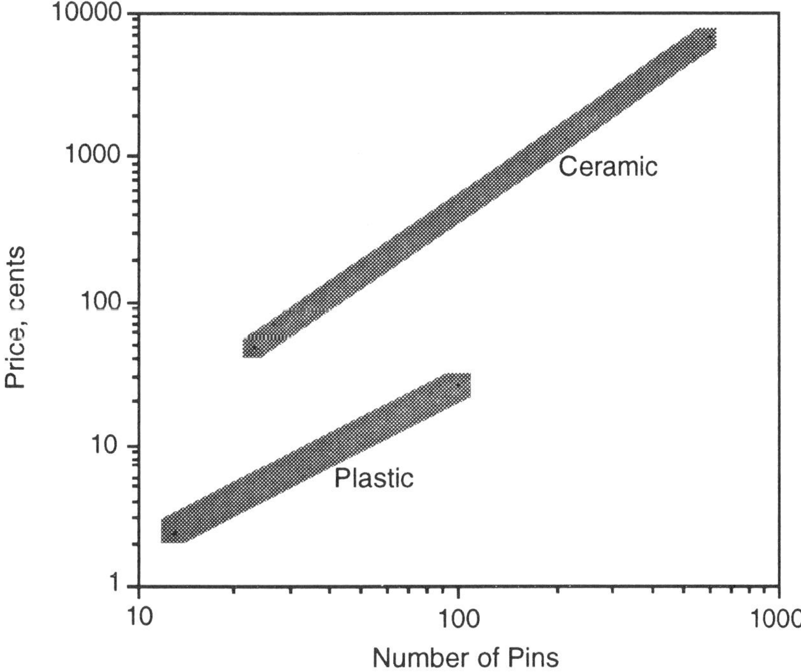

Figure 3.40. Price versus pin count assumptions for plastic and ceramic single chip packages, [3.70]. Not applicable for ball grid array packages (BGAs and OMPACs).

are not the result of a detailed derivation or process simulation, but rather a summary of quotes provided by interconnect suppliers.

Another measure of interconnect cost is the cost per wiring line length. This metric suffers from approximately the same problem as the previous one, it does not account for the fact that, because of smaller module area requirements, applications which make use of high density interconnects usually require a shorter total length of wiring. This metric is also affected by the placement and routing techniques used.

A more reasonable metric may be the cost per interconnect bond pad, [3.72]. This metric is more useful for a stand alone comparison of interconnects because it is the common top surface function that all interconnection technologies deliver. Cost per interconnect pad is unambiguous, can be readily counted, and does not rely on rule-of-thumb estimates or other subjective interpretations. A similar metric,

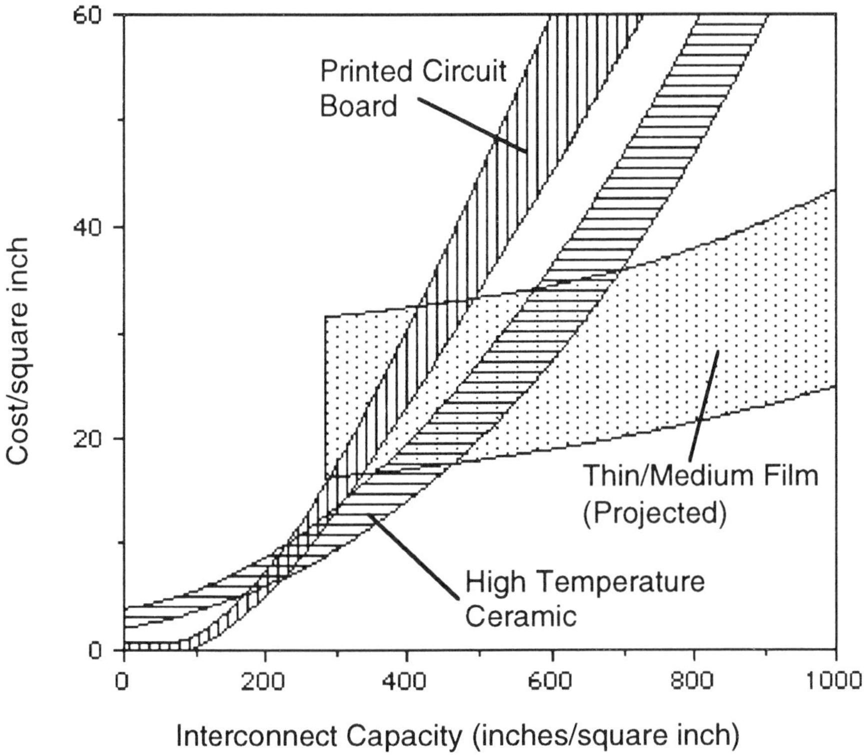

Figure 3.41. Cost per square inch of major interconnect technologies, [3.71]. (© 1992 IEEE)

suggested in [3.73] is the cost per device.

The metrics discussed above represent fair comparisons of interconnect costs if qualified with information about the application being packaged, however, in order to be accurate they must not be taken out of context.

Computational methods for interconnect costs have been developed and can be used. The computation of interconnect costs requires the characterization of detailed process steps. A review of detailed computational methods is provided in [3.74]. Detailed cost analysis can be broken into three categories: traditional cost analysis, activity-based cost analysis, and manufacturing simulation approaches. Traditional cost analysis computes the cost of a product based on the labor content and materials required. Burden or overhead is added as a percentage of the direct labor. This approach works well as long as the labor cost is the most significant cost driver. Activity-based cost analysis determines the cost of products based on the activities performed to create them. Activity-based analysis does not depend directly on the labor content of an activity but rather the number of times the activity is required and the number of changeovers in those activities. Unlike the first two approaches, simulation approaches are not accounting based. Simulation approaches provide the most accurate cost estimations at the expense of increased detail and complexity. Two approaches have been used. The most basic approach is to use a factory simulation where each process step is modeled and the process steps are then executed in the appropriate sequence simulating the actual manufacturing flow. The second approach is to use technical cost modeling (or process-based cost modeling). Technical cost modeling is an a priori cost model based on manufacturing simulation.

One of the factors which determines the cost of an interconnect for a particular technology is its size. The cost of a module may decrease as its area decreases due to better yields[6] and less material cost. For some interconnect technologies, the cost depends on the size and shape of the module due to changes in the "number up" (number of separate interconnects which can be fabricated on a single wafer or panel). The effective interconnect cost per cm^2 for different size interconnects is shown in Figure 3.42. A 12.5 cm diameter round wafer was used, all the modules were assumed to be square, and a processed wafer cost of $500 was assumed. Data on the number of interconnects which fit on this wafer was obtained from [3.75]. From Figure 3.42, it is clear that a small size difference can significantly impact the number of interconnects which can be fabricated on one wafer. Figure 3.42 also shows how the effective cost per cm^2 of the interconnects varies. Note that the effective cost per cm^2 of the interconnects does not continuously decrease as

6. Note that yield does not necessarily have to increase just because the module size is smaller. Decreasing the size of the module may be due to the use of finer lines or additional layers in which case the yield could drop as size decreases. If, however, the module can be decreased in size due to better utilization of the wiring resources or reduction in the number of passive components the yield would improve.

more modules are fabricated on the wafer, but rather saturates at one value. An obvious, but important, conclusion from this figure is that the interconnect size and shape can have a significant impact on the module cost when the number up is low. However, when the number-up is high, module size and shape will be less important.

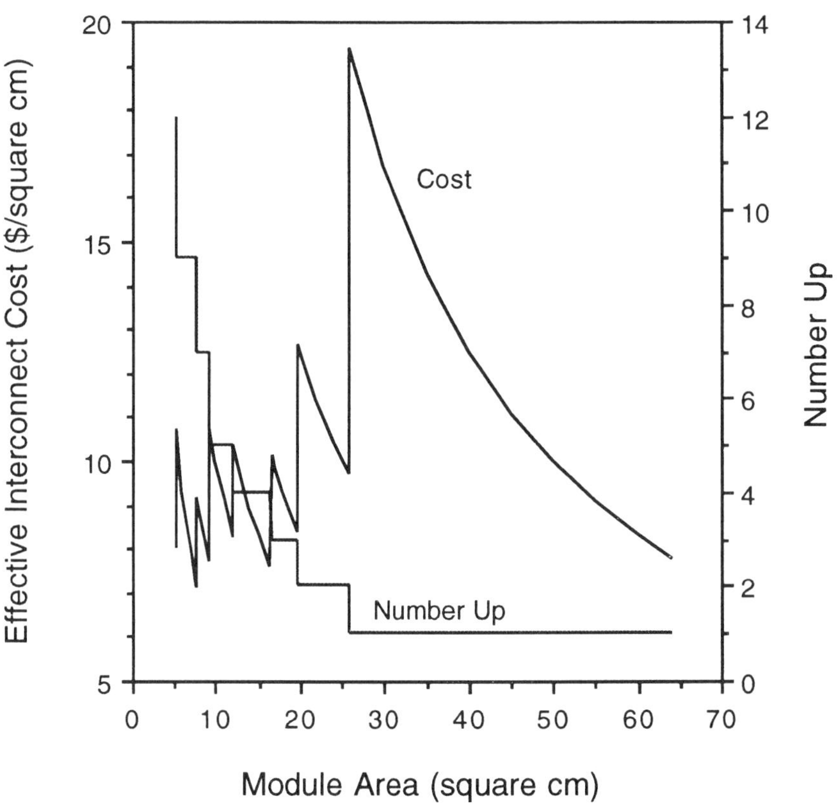

Figure 3.42. The effects of interconnect size on the number of interconnects which can be fabricated on one wafer (Number-Up) and the resulting effective (unyielded) cost per square cm.

Module Cost. The detailed costing of chips and interconnects is a necessary step in costing multichip systems, but by itself, it does not provide enough information to make accurate economic tradeoffs. Cost based tradeoffs can only be made by considering the completed module or system cost including the cost of chips, interconnects, assembly, test, and rework/repair. Assembled module costs are very complex to estimate. While considerable attention has been given to the costs of fabricating boards and interconnects, and the cost of various bonding methods, little comparative data exists on completed module costs. To complicate matters further, the results which have appeared are application specific, shedding only limited light on general trends.

To analyze the relationship among all the assembly, yield, test, and repair parameters, and how they impact the final cost and quality of a multichip system, a simple generic assembly, test, rework simulation model can be used. Figure 3.43 illustrates how the three basic process models are used to create a simple manufacturing flow.

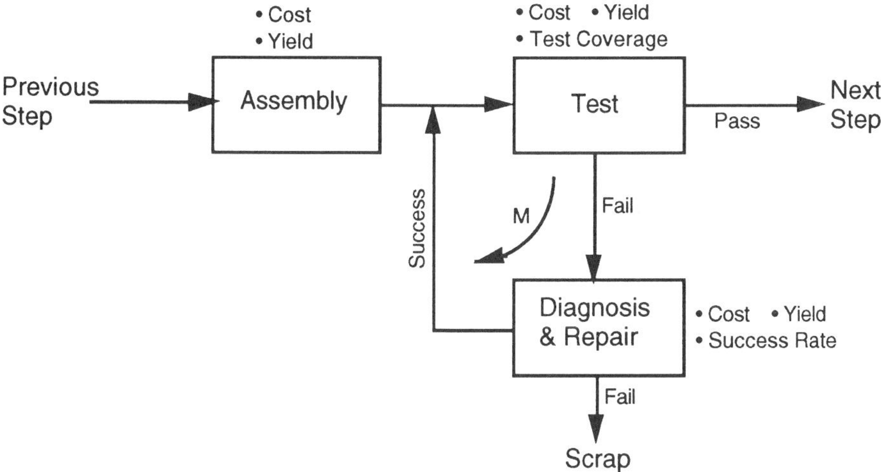

Figure 3.43. A simple multichip system manufacturing and test model [3.76]. M represents the maximum number of rework processes allowed.

Each of the boxes in Figure 3.43 represents a specific process object. The assembly box takes one or more inputs that correspond to various types of components to be assembled and the interconnect they are to be assembled upon. Each of the inputs is characterized by a cost, yield, and a count describing how many times it is used. The assembly process itself has a cost and a yield, as does the system which is produced. Equation (3.109) is used to compute the output cost of an assembly step,

$$cost_{output} = cost_{assembly} + \sum_{i=1}^{n} cost_{input}(i)\, count_{input}(i) \tag{3.109}$$

where n is the number of different input units being assembled in the assembly process (i.e. for 10 SRAM wirebonded to a interconnect and 1 VLSI device flip chip bonded to the same interconnect, n = 2, $count_{input}(1) = 10$, and $count_{input}(2) = 1$). $cost_{input}$ is the accumulated cost of the incoming components prior to the assembly step and $cost_{assembly}$ is the cost added by the assembly step. The output yield of the assembly step is given by,

$$yield_{output} = yield_{assembly} \prod_{i=1}^{n} (yield_{input}(i))^{count_{input}(i)} \tag{3.110}$$

The assembly object can represent one detailed process step such as dispensing die attach material or could represent a combination of many process steps such as a whole wirebonding process.

After assembly, the system or module is tested using a process that is characterized by a cost and fault coverage. The modules that pass the test are assumed good and are passed to the next activity (possibly another assembly step). The fraction of parts which pass (pass fraction(0)) is given by (3.107) where $yield_{wd}$ is replaced by $yield_{input}(0)$. Equation (3.107) does not, however, take into account the possibility of repairing bad units. The fraction of bad units detected (defect level) by the test is given by, 1 - pass fraction(0). Figure 3.44 shows the defect level of the module as a function of the module yield and the test coverage.

Since the testing process cannot detect all possible faults (test fault coverage < 100%), some defective modules escape and reduce the quality of the output modules. The fraction of escape units (relative to the total number of units which enter the test is given by

$$escapes(0) = pass\ fraction(0) - yield_{input} \tag{3.111}$$

The yield of units which exit the test (assuming no repair) is given by

$$\text{yield}_{\text{output}}(0) = \frac{\text{pass fraction}(0) + \text{escapes}(0)}{\text{pass fraction}(0)} \tag{3.112}$$

The output cost from the test step is given by,

$$\text{cost}_{\text{output}}(0) = \frac{\text{cost}_{\text{input}} + \text{cost}_{\text{test}}}{\text{pass fraction}(0)} \tag{3.113}$$

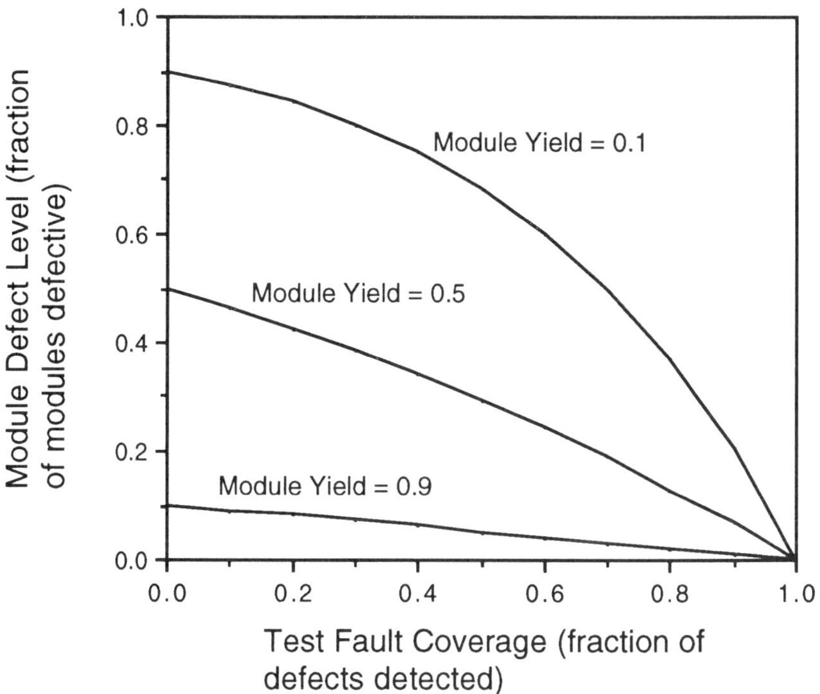

Figure 3.44. Module defect level versus test coverage. The module defect level is given by

$$\text{defect level} = 1 - \text{yield}_{\text{output}} = 1 - (\text{yield}_{\text{module}})^{(1 - \text{test coverage})}$$

where $\text{yield}_{\text{module}}$ is the module yield on the figure.

The modules that fail the test are then diagnosed for possible repair (some modules may not be repairable and are scrapped). The probability of being able to repair a unit is referred to as the repair success rate. Note that repairing a unit does not, in itself, guarantee that the unit is functioning properly (the unit may have been improperly diagnosed or new defects may have been introduced by the repair activity). The repairable modules are repaired and are again subjected to the testing process. The diagnosis and rework process has an average cost and also has a yield (i.e., not all repaired modules are good due to rework assembly defects, new component defects, accidental defects, or because of misdiagnosis). A module is scrapped if it fails to pass the test after being repaired a maximum allowable number of times (M). The following algorithm is used to compute the cost accumulated in this process,

$$\text{cost}_{\text{accumulated}} = \text{cost}_{\text{input}} + \text{cost}_{\text{test}} \tag{3.114}$$

repeat the following process for i = 1 to M repair cycles - for the ith repair cycle the following quantities are computed,

$$\text{scrap}(i) = (\text{bad fraction}(i-1))(\text{repair success rate}) \tag{3.115}$$

$$\text{cost}_{\text{accumulated}} = \text{cost}_{\text{accumulated}} + (\text{bad fraction}(i-1))\,\text{cost}_{\text{repair}} \tag{3.116}$$

$$\text{repaired}(i) = \text{bad fraction}(i-1) - \text{scrap}(i) \tag{3.117}$$

$$\text{bad fraction}(i) = \text{repaired}(i)\left(1 - (\text{yield}_{\text{repair}})^{\text{test coverage}}\right) \tag{3.118}$$

$$\text{cost}_{\text{accumulated}} = \text{cost}_{\text{accumulated}} + (\text{repaired}(i))\,\text{cost}_{\text{test}} \tag{3.119}$$

$$\text{pass fraction}(i) = \text{pass fraction}(i-1) + (\text{repaired}(i))(\text{yield}_{\text{repair}})^{\text{test coverage}} \tag{3.120}$$

$$\text{escapes}(i) = \text{escapes}(i-1) + \\ + (\text{repaired}(i))\left((\text{yield}_{repair})^{\text{test coverage}} - \text{yield}_{repair}\right) \tag{3.121}$$

The final parameters are computed from,

$$\text{cost}_{output} = \frac{\text{cost}_{accumulated}}{\text{pass fraction}(M)} \tag{3.122}$$

$$\text{yield}_{output} = \frac{\text{pass fraction}(M) - \text{escapes}(M)}{\text{pass fraction}(M)} \tag{3.123}$$

$$\text{total fraction scrapped} = \text{scrap}(M) + \text{bad fraction}(M) \tag{3.124}$$

where,

M = maximum number of repairs before scrapping
scrap(i) = the scrap fraction during repair cycle i
bad Fraction(i) = the fraction of bad units resulting from testing in cycle i
repaired(i) = fraction of units repaired in cycle i and not retested.

The simple three element model discussed above can be used to model larger portions of the manufacturing flow. By sequencing the appropriate type of processes, many different manufacturing flows can be constructed at virtually any level of desired detail and thereby optimization of the assembly/test/rework process is possible. Figure 3.45 shows a simplified process flow from wafer to finished MCM.

Known Good Die. One of the significant challenges which accompanies the manufacturing of multichip modules is the test and burn-in of the unpackaged active components. In traditional packaging approaches, all die are in their own single chip package, and therefore, can be readily tested and burned-in. Full functional tests can be performed by socketing the single chip packages and running them at the operating frequency and over the full temperature range. For die in single chip

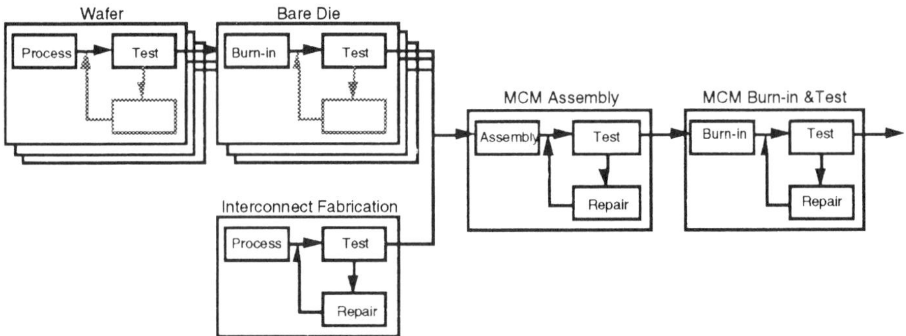

Figure 3.45. An extended test and manufacturing process flow model for an MCM.

packages, the as-received chip yield can be 0.99999 or higher. Unpackaged die do not have the same probability of being good since similar test and burn-in capabilities do not exist. Unpackaged die receive some testing at the wafer level (before sawing), however, these tests usually only check a limited number of parameters and are not performed at speed. Limited wafer testing usually results in as-received yields of 0.5 to 0.99 which are several orders of magnitude less than those for packaged die [3.77]. The relatively large expense of multichip module packaging is increased more when extensive rework and repair to replace defective active components is required because the components could not be adequately tested prior to module assembly.

The relationship between chip yield and first pass module yield is a simple function of the number of devices on the module, [3.75]. Figure 3.46 suggests that, in modules which contain relatively large numbers of devices with low yields, it may be cost effective to devote extra resources to the test and burn-in of the die before assembly. Alternatively, it may be economical to consider the use of TAB bonding or single chip packages over wirebonding and flip chip because of their test and burn-in advantages. There is a tradeoff between the cost of chip test before assembly and the cost of rework after. Depending on the application, it may be less expensive to repair the module than to perform extra testing of the chips (or add extensive built-in test to the chips). In other applications the cost of rework may be high enough to make extensive chip test and burn-in economical. For example, chip-first overlay interconnect approaches are extremely sensitive to chip yield since repairing a

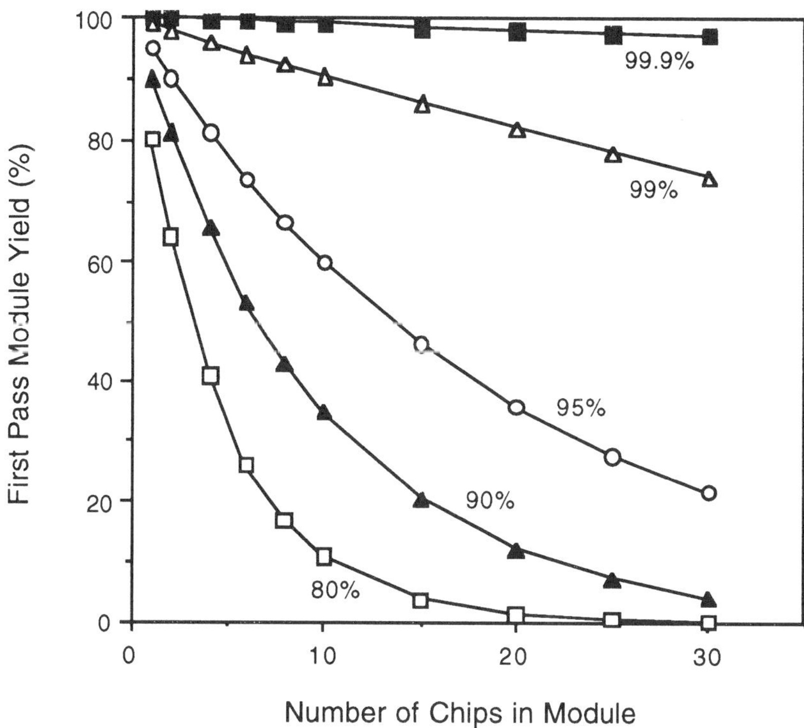

Figure 3.46. First pass module yield as a function of the number of chips in the module and the average chip yield at assembly (chip yield of 80% to 99.9% are shown on the figure). The module yield is given by

$$\text{Yield}_{\text{module}} = (\text{Yield}_{\text{chip}})^{N_{\text{chip}}}$$

where N_{chip} is the number of chips in the module.

completed module requires complete removal of the interconnect layers, thus making extensive chip testing prior to assembly, economical for many applications.

Another alternative to chip test or module rework is the addition of redundancy in the component set. Chip redundancy may be a viable alternative for applications which have ample space for extra chips (physical space and wiring resources) or contain a large number of identical chips (such as memory modules). Figure 3.47 shows the effect on first pass module yield of including one or two redundant chips in a module constructed with chips having an 80% yield at assembly. The figure shows that the first pass module yield can be doubled or tripled depending on the number of chips in the design.

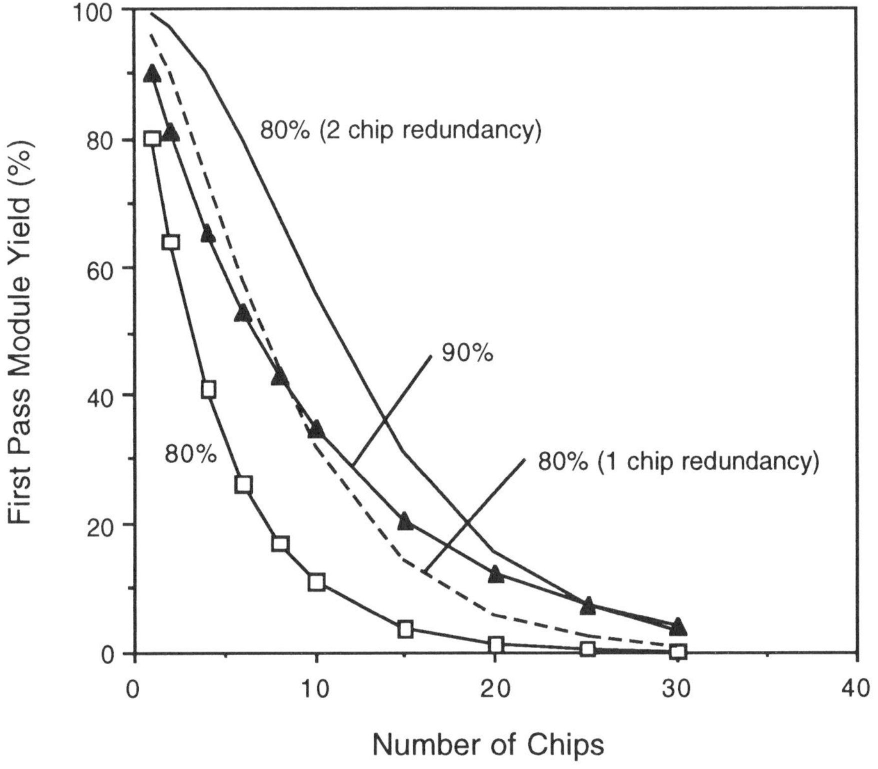

Figure 3.47. The effects of chip redundancy on first pass yield. While adding redundant chips to modules is not practical for all applications, it is an alternative to extensive chip test or rework problems. The module yield is given by

$$\text{Yield}_{\text{module}} = \sum_{i=N-K}^{N} \binom{N}{i} \text{Yield}^{i}_{\text{chip}} \left(1 - \text{Yield}_{\text{chip}}\right)^{N-i}$$

where N is the total number of chips and K is the number of redundant chips. The number of chips plotted on the horizontal axis in the figure is $N_{chip} = N - K$.

3.4 Tools for Tradeoff Analysis

In this section, we briefly review some examples of existing design advisor and design advisor manager tools. The design advisor and design advisor manager components of tradeoff analysis were previously outlined in Section 3.2 and are shown in Figure 3.2.

3.4.1 Design Advisors

The following paragraphs describe examples of tools which could provide design advisor assistance for a particular design view at a pre-physical design level (i.e., before a netlist or pin definitions exist). We have made no attempt to list all the applicable advisor tools. In many cases, applicable analysis modules are embedded within CAD tools performing other tasks. For example, constraint driven routing tools may contain analysis modules which are applicable in some cases to the conceptual design problem.

Several advisor tools for electrical modeling have been reported. Most of the tools provide simple advisor and management front ends to more detailed simulators. The PDQ (Pre-route Delay Quantifier), commercially available from Quad Design requires a component placement and module netlist as inputs and estimates the interconnect delays, line lengths, line loadings, and signal reflection. Optionally, the PDQ tool can generate an input file for a detailed transmission line simulator. The PDQ tool is designed to operate before routing of the module. PDDA (Power Distribution Design Advisor), developed at MCC [3.78], is a design advisor and model building tool specifically for the analysis of switching noise in single module/board systems. The tool allows the quantity and location of bypass capacitors to be determined through a "what if" approach. The tool is presently implemented on a Apple Macintosh using HyperCard and includes detailed simulation links to SPICE. A single chip package design advisor was developed at Intel [3.79]. This advisor performs electrical analysis of single and multilayer plastic and ceramic single chip packages via building equivalent circuit models for use in a more detailed circuit simulation tool. The advisor can be used to estimate package parasitics such as crosstalk, switching noise and package loading.

There are several reliability advisors which manage the metrology of failure rate manipulation, but do not necessarily compute the failure rates corresponding to the system elements itself. Applicable tools are not necessarily limited to the electronics domain. These tools allow the accommodation of various types of redundancy. A good example of this type of tool is Viable from Cadence Design Systems. Viable represents a general reliability platform which is independent of the methods used to predict the individual failure rates and allows users to add whatever methods they desire. MIL-HDBK-217 methods are available in the Viable tool for use as default failure rate predictors. Other tools that predict reliability based MIL-HDBK-217 have been developed, i.e., Lambda from CMU [3.80], and many commercial offerings. Tools focused on estimating dependability metrics for multichip systems (in the architectural design sense - not the packaging and interconnect sense) are discussed in [3.60] and [3.81]. One of these tools has been integrated into a multidisciplinary tradeoff tool. The ASSURE (Automated design for dependability) tool developed at Carnegie Mellon University, [3.60] and [3.82], is being used within the MICON system discussed in Section 3.4.2. ASSURE includes dependability analysis, evaluation of dependability enhancement techniques using predictive estimation, and the selection of a dependability technique.

Reliability advisors which are not MIL-HDBK-217 based are also being developed. The most notable effort is within the University of Maryland's CALCE (Computer Aided Life Cycle Engineering) program [3.83]. CALCE is developing design-for-reliability advisors for packaging systems based on a physics-of-failure approach.

Several manufacturability analysis advisors have been developed for traditional assembly approaches (through-hole and surface mounting). An example of this type of advisor tool has been developed at SUNY-Binghamton [3.84]. This tool is a rule-based tool for advising user on the manufacturability of surface mount printed circuit boards. IPEX (Integrated Product Engineering eXpert system) [3.85] developed at Westinghouse Electric Corporation, assists a design engineer in developing a product which is producible, supportable, and compliant to customer specifications. It provides an engineer with a means of documenting and searching guidelines applicable to a product and allows an evaluation based on those guidelines to be made. IPEX is specific to the development of printed circuit board assemblies.

A large number of commercially available advisor tools exist for thermal and cost analysis. In the case of thermal analysis, most two-dimensional thermal simulation tools are applicable to conceptual design.

For cost analysis, there are a few packaging specific tools available now. One such tool is from IBIS Associates. The IBIS tool performs technical cost modeling focused on MCM interconnects. See [3.74] for more discussion of this tool and its approach.

Extraction and Constraint Driven Routing Tools. Extraction tools analyze the performance of a module after or concurrent with the physical design activities of routing, placement, and layout. All major CAD vendors have developed extraction tools which can be linked with physical design processes. For example, the "Signal Noise Analysis" application in Cadence's Allegro-MCM accepts a module routing and the details of the net driver and receiver as input, and estimates the contributions to the nets' noise budget from reflections, crosstalk, thermal shifts, and ohmic losses. Mentor's MCM Station tool set also offers integration of thermal and signal analysis with placement and routing. Mentor's signal analysis approach differs from that of Cadence in the sense that Mentor used Quad Design's crosstalk toolkit (an electromagnetic simulation package) instead of closed form analytical formulations. Other extraction tools estimate electrical delays, timing, reliability, and link to thermal or electrical simulators. Extraction and performance driven layout do not have a direct impact on tradeoff analysis, however, in some cases the estimations they use are applicable to the conceptual design level.

If physical layout and detailed analysis are used concurrently in an automated fashion, the combination is often referred to as "performance driven layout" or "force-directed placement". The two primary drivers in performance driven layout tools which have appeared to date are timing (delay) [3.86] and reliability (thermal) [3.87]. Performance driven layout uses performance requirements as additional direct constraints during layout and routing. Timing driven layout uses the criticality measure of nets to bias placement and routing programs, and thereby, can bring the performance of a chip significantly closer to that of an ideal layout. Reliability driven layout couples reliability and routability based placement with the objective of improving the total reliability of all components while minimizing the total wire length. SURF (Santa-Cruz ULSI Routing Framework) from the University of California Santa Cruz combines performance driven layout and routing taking into account electrical and thermal constraints for MCMs, [3.88].

Many commercial packages for thermal/reliability/electrical analysis exist for "what if layout". For example, with ViewPlace from Viewlogic Systems Inc., a designer can perform a thermal analysis on a design and then automatically invoke Reap, a reliability-prediction tool from Systems Effectiveness Associates Inc. Reap calculates reliability figures based on standard reliability models. Designers can work interactively with the ViewPlace/Reap combination to optimize a design's reliability. Fastrace from Algorex Corp. simulates the behavior of laid-out printed circuit boards and applies controlled stub routing constraints to reduce total wiring lengths.

3.4.2 Design Advisor Managers

The following tools each attempt to concurrently treat a subset of the interdisciplinary system synthesis/specification space using a mixture of estimation, simulation, and synthesis techniques. What sets these tools apart from constraint driven routing and extraction based physical design tools is that they all are capable of operating at a pre-physical design level (i.e., before a netlist or pin definitions exist).

SUSPENS (Stanford University System PErformaNce Simulator). SUSPENS [3.9, 3.89, 3.90] was developed at Stanford University as a "system-level circuit model for central processing units." SUSPENS predicts clock frequency, power dissipation, number of chip I/O, and chip size for silicon NMOS, CMOS, and BJT, and GaAs MESFET, HEMT, and HBT. SUSPENS begins with the number of gates per chip and collects inputs such as the technology type and minimum feature size, then provides interconnection pitches, number of wiring layers, fanout of gates, efficiency of interconnections, and parasitic capacitance and resistances from databases. SUSPENS uses a Donath style Rent's rule estimation for net lengths (and thereby module sizes for homogeneous component sets). SUSPENS has been extensively tested against real microprocessors, gate arrays, and mini- and mainframe computers. The SUSPENS tool is structured such that the analytical expressions characterizing a specific technology can be easily replaced with libraries of models for specific circuit families.

AUDiT (AUtomated Design and Tradeoff simulator for integration of computer structures). AUDiT [3.91, 3.92] was developed at Cornell University. AUDiT represents a more detailed version of the SUSPENS work with extensions into module and multiple module estimations. Like SUSPENS, AUDiT starts with a circuit count and/or bit count along with a technology database that includes information about each logic type and feature size dictated design rules. AUDiT estimates the number of I/O per component, component dimensions, on-chip electrical properties including crosstalk and delta-I noise, on-chip delay quantities, and power dissipation. AUDiT's module-level abilities include delay estimation (and associated metrics: clock rate, cycle time, MIPS), module size, and power dissipation. The delay and timing estimations in AUDiT are considerably more sophisticated than those in the earlier SUSPENS tool. AUDiT includes a simulated annealing numerical optimizer.

MICON (MIcro-processor CONfigurer). MICON [3.93, 3.94, 3.95] was developed at Carnegie Mellon University. MICON differs from SUSPENS, AUDiT, and the MSDA tool (to be discussed later in the next section) in the sense that it is netlist based versus physically based, i.e., the goal of MICON is to select an appropriate set of components and generate a netlist for the module using power, area, and cost constraints. MICON is a library-based chip set synthesis system. MICON does not have the breath of physical definition or technology knowledge in the packaging area which the SUSPENS, AUDiT, and MSDA tools have. However, the netlist orientation of MICON and its behavioral level interface bring it closer to being a "system compiler" than any of the other tools. The MICON tool is being commercialized by Omniview, Inc. The commercial version of the tool is called Fidelity. Fidelity provides a VHDL link to Synopsis for ASIC synthesis. Omniview is developing several advisor modules including design-for-test and design-for-reliability for Fidelity. NTT has reportedly developed a similar system based on the MICON work [3.96].

PEPPER. PEPPER [3.13] was developed at IBM. PEPPER focuses on functional partitioning, chip placement, package congestion, pin assignment, and coupled noise analysis. PEPPER requires a technology definition, wiring strategy choice, and a design description as inputs, and performs a net topology analysis (routing pre-screen) and a congestion analysis. A more accurate description of the PEPPER tool is as a constraint driven pre-router (the constraints being delay and noise). PEPPER can provide significantly more accurate routing, chip placement, and layer requirement predictions than any of the other tradeoff tools discussed here, however, PEPPER requires significantly more inputs than the other tools including: net-topology constraints and wiring rules. PEPPER has a built in optimization capability like AUDiT.

Yoda. Yoda [3.1] was developed at Carnegie Mellon University. Yoda is a VLSI system planning tool targeted at the domain of digital signal processing and in particular the task of designing digital filters. Yoda starts with the functional specifications of the desired frequency response and allows the designer to examine a wide variety of possible filter design plans. Yoda works almost exclusively at the chip-level and may not seem to fit with the other five tradeoff tools discussed above and in the next section, however, the philosophy developed within the Yoda work is closer to the approach discussed in this book than any of the tools discussed above.

One additional tool exists which encompasses the ideas of a design advisor manager targeted at tradeoff analysis in multichip systems. Many of the algorithms discussed in this book are implemented in the tool discussed in the next section.

3.4.3 The MSDA (Multichip Systems Design Advisor) Tool

MSDA [3.26, 3.97] has been developed in a consortia funded project at MCC. MSDA is a software tool for enhancing the manufacturability and decreasing the design risk associated with the selection of packaging technologies for integrated circuits. MSDA[7] is designed to concurrently compute physical (size, weight, interconnect routing requirements), electrical (delays, noise), thermal (internal and external thermal resistances), reliability (MTTF), testability, and cost performance metrics for multichip systems. Technologies treated by MSDA

7. Also referred to as SPEC (System Performance Evaluation and Comparison) in some published references.

include: traditional and fine-line printed circuit boards, low temperature co-fired ceramic, and thin-film (chip-first and chip-last). Multiple design rule sets for each of these technologies are included in the tool, however the user may easily customize the properties or add their own proprietary technologies. Component assembly approaches include wirebonding, TAB, flip chip, and single chip packages. Materials are also available for bare die attach, encapsulation, heat exchanger attach, and for defining the bonding and interconnect technologies.

Figure 3.48 shows a block diagram of the MSDA tool. The tool begins by capturing design and performance budgets and constraints. Next the components (active and passive) to be packaged are described. Component descriptions include the entry of physical (dimensions), material, I/O (number of signal, power, ground, unused), electrical (bias level, logic swing, simple driver and receiver descriptions or driver and receiver SPICE models), thermal (allowable junction temperature, power dissipation), and manufacturing (cost, tested yield, burned-in yield, test and burn-in costs). All of this information is included in the "chip" class. The chip class also contains details of the type of bonding, die attach material, encapsulation, and single chip package (if there is one) associated with a particular component. In addition to these properties, the characteristics of heat spreaders and thermal vias are captured as properties of each component rather than the module or board so that they can be easily manipulated on a per component basis.

After the components are described to the tool, they are placed into "partitions." Partitions consist of any subset of the defined components grouped into a few-chip package, MCM, or printed circuit board. It is through the creation of partitions that complex systems can be built using the tool. The characteristics of partitions include interconnect technologies, heat exchangers, and connectors. A netlist corresponding to the partition under design, a group of partitions, or a design larger than that under consideration is an optional input to the tradeoff tool, however, often conceptual design takes place prior to the synthesis of a netlist. The partitions are designed to be hierarchical and organized in a tree structure so that they may be placed inside of each other to build systems and study tradeoffs associated with alternative distributions of components amongst system interconnects. The hierarchical nature of partitions is possible because an integrated circuit and a completed module share many identical properties and can often be treated in the design of a system in a similar manner.

Besides providing concurrent performance estimation through closed-form algorithms, the MSDA tool supplements its internal analysis by linking to detailed simulation through model building and simulator management activities. Detailed simulation links are presently used as part of the electrical performance analysis and reliability evaluation in the tool.

The analysis of critical net delays is one example of the fusing of es-

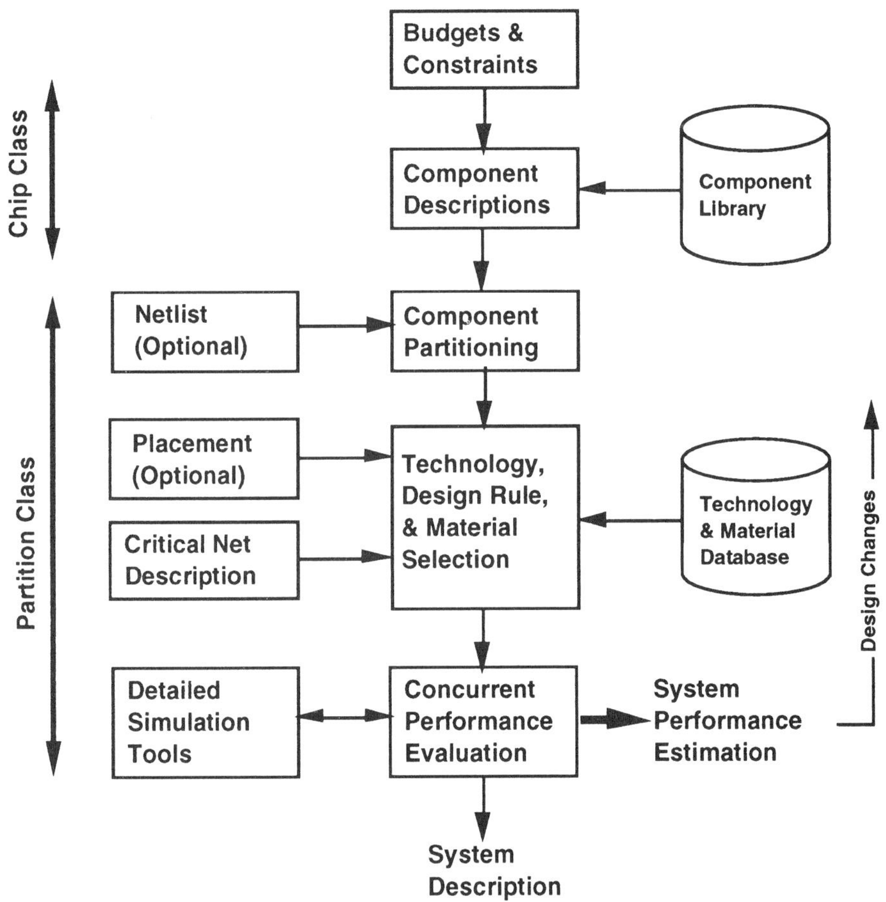

Figure 3.48. Block diagram of the MSDA tradeoff analysis tool [3.26]. (© 1993 IEEE)

timation level analysis and simulation, and the effective use of simulators in conceptual design activities. The evaluation of critical net delays at the conceptual design level consists of the formulation of a "packaging delay" which includes time-of-flight, RC charging, and contributions from reflections. Packaging delays along with noise contributions, and the delays associated with gates, receivers and memory access times determine the maximum operating frequency of multichip systems.

The process by which critical net delays are evaluated in the tradeoff tool is shown in Figure 3.49. To evaluate delays in the MSDA tool, the designer draws a critical connection on a placement editor used to capture conceptual level layout information. The net may have multiple loads, bends, vias, and stubs. Next the net path and connections are pro-

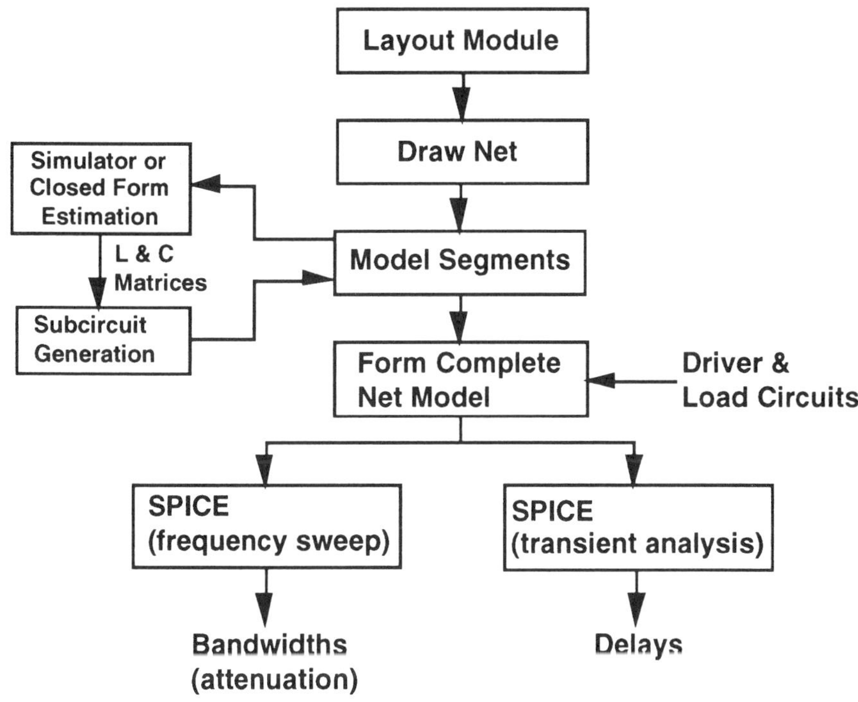

Figure 3.49. Critical net delay and bandwidth (attenuation) evaluation. A segment is defined in the model as the length of line between two nodes (i.e. no internal bends, taps, or vias). An example output of the process is shown in Figure 5.3 in Chapter 5 [3.26]. (© 1993 IEEE)

cessed by the tool to form a model of each net segment for use in a detailed electromagnetic simulator. The simulator computes inductance and capacitance matrices for the segment which are post processed by MSDA into SPICE subcircuits. After all the segments are modeled, the subcircuits are combined along with driver and receiver models (either input with the component descriptions in subcircuit form or defined using Thevenin equivalent circuits from user supplied characteristics) and the whole net is analyzed using SPICE. All of the simulation management and post processing is handled by the tradeoff tool and performed concurrently with size, routing, cost, reliability, testability, and thermal analyses.

3.5 References

[3.1] A. M. Dewey and S. W. Director, *Principles of VLSI System Planning: A Framework for Conceptual Design*, Kluwer Academic Publishers, 1990.

[3.2] J. L. Sloan, *Design and Packaging of Electronic Equipment*, Van Nostrand Reinhold Company, 1985.

[3.3] B.S. Landman and R.L. Russo, "On a Pin Versus Block Relationship for Partitions of Logic Graphs," *IEEE Transactions on Computers*, vol. C-20, pp. 1469-1479, December, 1971.

[3.4] W. Donath, "Equivalence of Memory to 'Random Logic'," *IBM Journal of Research and Development*, vol. 58, no. 5, pp. 401-407, September, 1974.

[3.5] M. Feuer, "Connectivity of Random Logic," *IEEE Transactions on Computers*, vol. C-31, pp. 29-33, January, 1982.

[3.6] I. E. Sutherland and D. Oestreicher, "How Big Should a Printed Circuit Board Be?," *IEEE Transactions on Computers*, vol C-22, no. 5, pp. 537-542, May, 1973.

[3.7] Seraphim, D.P., "Chip-Module-Package Interfaces," *Proceedings of Electronic Insulation Conference*, pp. 90-93, September, 1977.

[3.8] W. E. Donath, "Placement and Average Interconnection Lengths of Computer Logic," *IEEE Transactions on Circuits and Systems*, vol CAS-26, no. 4, pp 272-277, April, 1974.

[3.9] H. B. Bakoglu, *Circuits, Interconnections, and Packaging for VLSI*, Addison Wesley, 1990.

[3.10] R. Hannemann, "Physical Technology for VLSI Systems," *Proceedings of the International Conference on Computer Design*, pp. 48-53, 1986.

[3.11] L. L. Moresco, "Electronic System Packaging: The Search for Manufacturing the Optimum in a Sea of Constraints," *IEEE Transactions on Components, Hybrids, and Manufacturing Technology*, vol. 13, pp. 494-508, September, 1990.

[3.12] D. C. Schmidt, "Circuit Pack Parameter Estimation Using Rent's Rule," *IEEE Transactions on Computer-Aided Design of Integrated Circuits and Systems*, vol. CAD-1, no. 4, pp. 186-192, October, 1982.

[3.13] D. P. LaPotin, T. R. Mazzawy, and M. L. White, "Early Package Analysis: Considerations and Case Study," *IEEE Computer*, vol. 26, no. 4, pp. 30-39, April, 1993.

[3.14] B. Freyman and R. Pennisi, "Over Molded Pad Array Carrier Package - OMPAC," *Proceedings of the 41st Electronic Components and Technology Conference*, pp. 176-185, 1991.

[3.15] T. S. Steele, "Terminal and Cooling Requirements for LSI Packages," *IEEE Transactions on Components, Hybrids, and Manufacturing Technology*, vol. 4, pp. 187-191, June, 1981.

[3.16] P. A. Sandborn, "Technology Application Tradeoff Studies in Multichip Systems," *Proceedings of the 1st International Conference on Multichip Modules*, pp. 150-158, 1992.

[3.17] V. W. Antonetti, S. Oktay, and R. E. Simons, "Heat Transfer in Electronic Packages," *in Microelectronics Packaging Handbook*, R. R. Tummala and E. J. Rymaszewski editors, Van Nostrand Reinhold, pp. 167-223, 1989.

[3.18] D. P. Kennedy, "Heat Conduction in a Homogeneous Solid Cylinder of Isotropic Media," *Journal of Applied Physics*, vol. 31, no. 8, pp. 490-497, August, 1960.

[3.19] M. M. Yovanovich, "Thermal Contact Resistance Course Notes," *Presented at the 1989 International Electronics Packaging Conference (IEPS)*, September, 1989.

[3.20] T. Dolbear, "Thermal Management of Multichip Modules," *Electronic Packaging and Production*, vol. 32, no. 6, pp. 60-63, June, 1992.

[3.21] D. Q. Kern and A. D. Kraus, *Extended Surface Heat Transfer*, McGraw-Hill Book Company, 1972.

[3.22] R. J. Phillips, "Forced-Convection, Liquid-Cooled, Microchannel Heat Sinks," *Technical Report 787*, Lincoln Laboratory, Massachusetts Institute of Technology, January, 1988.

[3.23] D. B. Tuckerman, "Heat-Transfer Microstructures for Integrated Circuits," Ph.D. Dissertation, Stanford University, February, 1984.

[3.24] "The NBS-NACA Tables of the Thermal Properties of Gases, NBS ref. W-4391, S50-35.

[3.25] A. D. Kraus and A. Bar-Cohen, *Thermal Analysis and Control of Electronic Equipment*, McGraw-Hill Book Company, 1983.

[3.26] P.A. Sandborn, K. Drake, and R. Ghosh, "Computer Aided Conceptual Design of Multichip Systems," *Proceedings of the Custom Integrated Circuits Conference*, pp. 29.4.1-29.4.4, 1993.

[3.27] P. Penfield and J. Rubinstein, "Signal Delay in RC Tree Networks," *Proceedings of the 18th Design Automation Conference*, pp. 613-617, 1981.

[3.28] A. Wilnai, "Open-Ended RC Line Model Predicts MOSFET IC Response," *Electronic Design News*, vol. 16, pp. 53-54, 1971.

[3.29] K. C. Gupta, R. Garg, and R. Chadha, *Computer-Aided Design of Microwave Circuits*, Artech House, Inc., 1981.

[3.30] E. E. Davidson, "The Performance Choice: A Study of Chip and Package Densities," *Proceedings of the 40th Electronic Components and Technology Conference*, pp. 147-151, 1990.

[3.31] G. A. Katopis, "Electrical Interconnection Design for High Performance Machines," *Proceedings of the National Electronic Packaging and Production Conference (NEPCON-West)*, pp. 169-177, 1990.

[3.32] D. Balderes and M. White, "Packaging Effects on CPU Performance of Large Commercial Processors," *Proceedings of the 35th Electronic Components Conference*, pp. 351-356, 1985.

[3.33] V. K. Nagesh, D. Miller, and L. Moresco, "A Comparative Study of Interconnect Technologies," *Proceedings of the International Electronics Packaging Symposium (IEPS)*, pp. 433-443, 1989.

[3.34] R. Kaw, "Comparison of Chip Crossing Delay in Various Packaging Environments," *Proceedings of the IEEE International Conference on Computer Design: VLSI in Computers*, pp. 233-236, 1989.

[3.35] T. R. Gheewala, "System Level Comparison of High Speed Technologies," *Proceedings of the IEEE International Conference on Computer Design: VLSI in Computers*, pp. 245-250, 1984.

[3.36] IEEE Split Rock Workshop, 1985.

[3.37] R. Hannemann, "Interconnects and Packaging for Highly Integrated Systems," *Presented at SPIE International Conference on Advances in Interconnect and Packaging*, 1990.

[3.38] D. L. Carter and D. F. Guise, "Analysis of Signal Propagation Delays and Chip Level Performance Due to On-Chip Interconnections," *Proceedings of the IEEE International Conference on Computer Design: VLSI in Computers*, pp. 218-221, 1983.

[3.39] R. W. Keyes, "A Figure of Merit for IC Packaging," *IEEE Journal of Solid State Circuits*, vol. 13, no. 2, pp. 265-266, February 1978.

[3.40] *ITT, Reference Data for Radio Engineers*, 5th edition, Howard W. Sams & Co., 1968.

[3.41] G. A. Katopis, "Methodology and Computer Aids for the Noise Tolerance of ECL Logic Gates," *Proceedings of the 12th Asilomar Conference on Circuits, Systems and Computers*, pp. 500-505, 1979.

[3.42] W. R. Blood, *MECL System Design Handbook*, Motorola Inc., 1988.

[3.43] I. Catt, "Crosstalk (Noise) in Digital Systems," *IEEE Transactions on Electronic Computers*, vol. EC-16, pp. 743-763, 1967.

[3.44] P. M. Duncan, D. L. Fett, A. H. Mones, and T. R. Poulin, "Noise Characteristics on High Density TAB Configurations," *Proceedings of the National Electronic Packaging and Production Conference (NEPCON-West)*, pp. 638-649, 1990.

[3.45] A. J. Rainal, "Computing Inductive Noise of Chip Packages," *AT&T Bell Laboratories Technical Journal*, vol. 63, pp. 177-195, 1984.

[3.46] R. Kaw, R. Liu, K. Lee, and R. Crawford, "Effective Inductance for Switching Noise in Single Chip Packages," *Proceedings of the Tenth Annual International Electronics Packaging Conference*, pp. 756-761, 1990.

[3.47] A. T. Baba, W. D. Carlomagno, D. E. Cummings, and F. Guerrero, "Bonded Interconnect Pin (BIP) Technology, A Bare Chip Connection Process for Multichip Modules," *Proceedings National Electronic Packaging and Production Conference (NEPCON-West)*, pp. 1167-1177, 1991.

[3.48] P. Robock, and L. T. Nguyen, "Plastic Packaging," *in Microelectronic Packaging Handbook*, R.R. Tummala and E.J. Rymaszewski editors, Van Norstrand Reinhold, pp. 523-672, 1989.

[3.49] N. G. Koopman, T. C. Reiley, and P. A. Totta, "Chip-to-Package Interconnections," *in Microelectronic Packaging Handbook*, R.R. Tummala and E.J. Rymaszewski editors, Van Norstrand Reinhold, pp. 361-453, 1989.

[3.50] E. E. Davidson, "Electrical Design of a High Speed Computer Packaging System," *IEEE Transactions of Components, Hybrids, and Manufacturing Technology*, vol. CHMT-6, pp. 272-282, 1983.

[3.51] G. A. Katopis, "Delta-I Noise Specification for a High-Performance Computing Machine," *Proceedings of the IEEE*, vol. 73, pp. 1405-1415, 1985.

[3.52] P. A. Sandborn, H. Hashemi, and B. Weigler, "Switching Noise in a Medium Film Copper/Polyimide Multichip Module," *Proceedings of the SPIE International Conference on Advances in Interconnection and Packaging*, SPIE vol. 1390, pp. 177-186, 1990.

[3.53] H. Hashemi, P. A. Sandborn, D. Disko, and R. Evans, R., "The Close Attached Capacitor: A Solution to Switching Noise Problems," *IEEE Transactions on Components, Hybrids, and Manufacturing Technology*, vol. 15, pp. 1056-1063, December, 1992.

[3.54] R. Senthinathan, G. Tubbs, and M. Schuelein, "Negative Feedback Influence on Simultaneously Switching CMOS Outputs," *Proceedings of the Custom Integrated Circuits Conference*, pp. 5.4.1-5.4.5, 1988.

[3.55] R. F. Pease, B. W. Langley, and O. K. Kwon, "Simultaneous Switching Noise in Multichip Packaging," *SRC Technical Report T89067*, Stanford University, pp. 29-38, 1989.

[3.56] C. M. Val and J. E. Martin, "A New Chip Carrier for High Performance Applications Integrated Decoupling Capacitor Chip Carrier 'IDCCC'," *Proceedings of the International Electronics Packaging Conference*, pp. 143-159, 1988.

[3.57] H. Hashemi, U. Ghoshal, K. Ziai, and P. A. Sandborn, "Analytical and Simulation Study of Switching Noise in CMOS Circuits," *Proceedings of the International Electronics Packaging Conference*, pp. 762-773, 1990.

[3.58] R. E. Canright, "A Formula to Model Delta-I Noise," *Proceedings of the 37th Electronic Components Conference*, pp. 354-361, 1987.

[3.59] J. C. Laprie and A. Costes, "Dependability: A Unifying Concept for Reliable Computing," *FTCS-12 Digest of Papers*, pp. 18-21, 1982.

[3.60] P. Edmond, A. P. Gupta, D. P. Siewiorek, and A. A. Brennan, "ASSURE: Automated Design for Dependability," *Proceedings of the 27th Design Automation Conference*, pp. 555-560, 1990.

[3.61] D. C. Boose and J. M. Hsiao, "Model for Transient and Permanent Error Detection and Fault-Isolation Coverage," *IBM Journal of Research and Development*, vol. 26, no. 1, pp. 67-77, January, 1982.

[3.62] U.S. Department of Defense, Rome Air Development Center, "Reliability Prediction of Electronic Equipment, MIL-HDBK-217F," March, 1991.

[3.63] R. J. Allen and W. J. Roesch, "Reliability Prediction: The Applicability of High Temperature Testing," *Solid State Technology*, vol. 33, pp. 103-108, September, 1990.

[3.64] P. A. Tobias and D. C. Trinidade, *Applied Reliability*, Van Nostrand Reinhold Company, 1986.

[3.65] M. Pecht, M. Dasgupta, D. Barker, and C. Leonard, "A Reliability Physics Approach to Failure Prediction Modeling," *Quality and Reliability Engineering International*, vol. 6, no. 4, pp. 267-273, September-October, 1990.

[3.66] B. T. Murphy, "Cost-Size Optima in Monolithic Integrated Circuits," *Proceedings of the IEEE*, vol. 52, pp 1537-1545, December, 1964.

[3.67] R. B. Seeds, "Yield, Economic, and Logistic Models for Complex Digitial Arrays," *Proceedings of the IEEE International Convention Record, Part 6*, pp. 60-61, 1967.

[3.68] T. W. Williams and N. C. Brown, "Defect Level as a Function of Fault Coverage," *IEEE Transactions on Computers*, vol. C-30, no. 12, pp. 987-988, December, 1981.

[3.69] V. D. Agrawal, S. C. Seth, and P. Agrawal, "Fault Coverage Requirement in Production Testing of LSI Circuits," *IEEE Journal of Solid State Circuits*, vol. SC-17, pp. 57-61, February, 1982.

[3.70] G. Messner, "Price/Density Tradeoffs of Multichip Modules," *Proceedings of the International Symposium on Hybrid Microelectronics (ISHM)*, pp. 28-36, 1988.

[3.71] J. W. Balde, "Crisis in Technology: The Questionable U.S. Ability to Manufacture Thin Film Multichip Modules," *Proceedings of the IEEE*, vol. 80, no. 12, pp. 1995-2002, December, 1992.

[3.72] C. Lassen, "Integrating Multi-Chip Modules into Electronic Equipment: The Technical and Commercial Trade-Offs," *Proceedings of the Tenth Annual International Electronics Packaging Conference*, pp. 3-15, 1990.

[3.73] D. G. Kelemen, "Cost Considerations in High-Density Packaging," *Proceedings of the Ninth Annual International Electronics Packaging Conference*, pp. 1216-1222, 1989.

[3.74] L. H. Ng, "MCM Package Selection: Cost Issues," *in Multichip Module Technologies and Alternatives - The Basics*, D. A. Doane and P. Franzon editors, Van Nostrand Reinhold, pp. 133-164, 1993.

[3.75] G. R. Weihe, "High Density Multichip Solutions," *Proceedings of the National Electronic Packaging and Production Conference (NEPCON-West)*, pp. 1121-1129, 1991.

[3.76] M. Abadir, A. Parikh, L. Bal, P. Sandborn, and C. Murphy, "High Level Test Economics Advisor (Hi-TEA)," *Proceedings of the International Workshop on the Economics of Design, Test, and Manufacturing*, 1993.

[3.77] J. K. Hagge and R. J. Wagner, "High-Yield Assembly of Multichip Modules Through Known-Good IC's and Effective Test Strategies," *Proceedings of the IEEE*, vol. 80, no. 12, pp. 1965-1994, December, 1992.

[3.78] P. A. Sandborn and H. Hashemi, "A Design Advisor and Model Building Tool for the Analysis of Switching Noise is Multichip Modules," *Proceedings of the International Symposium on Microelectronics (ISHM)*, pp. 652-657, 1990.

[3.79] G. Choksi, B. K. Bhattacharyya, S. Stys, and B. Natarajan, "Computer Aided Electrical Modeling of VLSI Packages," *Proceedings of the 40th Electronic Components and Technology Conference*, pp. 169-172, 1990.

[3.80] S. A. Elkind, Lamda; *A MIL-217D System Failure Rate Analysis Program User's Manual*, ECE Department, Carnegie Mellon University, 1983.

[3.81] A. M. Johnson Jr. and M. Malek, "Survey of Software Tools for Evaluating Reliability, Availability, and Serviceability," *ACM Computing Surveys*, pp. 227-269, 1988.

[3.82] A. A. Brennan, "Automatic Synthesis for Reliability," *Research Report No. CMUCAD-88-3*, Carnegie Mellon University, January, 1988.

[3.83] "Computer Aided Design of IC, Hybrid, and MCM Packages - CADMP," *CALCE Electronic Packaging Research Center*, University of Maryland, 1992.

[3.84] K. Srihari and C. R. Emerson, "Knowledge Based Manufacturing Problem Diagnosis in PCB Assembly," *Proceedings of the Twelfth Annual International Electronics Packaging Conference*, pp. 245-257, 1992.

[3.85] D. L. Hilbert, "The Integrated Product Engineering Expert System (IPEX) A Design/Producibility Advisor," *Proceedings of the National Electronic Packaging and Production Conference (NEPCON-West)*, pp. 157-163, 1992.

[3.86] Y. Ogawa, T. Itoh, Y. Miki, T. Ishii, Y. Sato, and R. Toyoshima, "Timing- and Constraint-Oriented Placement for Interconnected LSIs in Mainframe Design," *Proceedings of the 28th Design Automation Conference*, pp. 253-258, 1991.

[3.87] M. D. Osterman and M. Pecht, "Placement for Reliability and Routability of Convectively Cooled PWB's," *IEEE Transactions on Computer-Aided Design*, vol. 9, pp. 734-744, 1990.

[3.88] W. W.-M. Dai, "Performance Driven Layout of Thin Film Multichip Modules," *Proceedings of the SRC Topical Research Conference on Packaging Design, Analysis, and Simulation*, pp. 35-37, 1991.

[3.89] H. B. Bakoglu and J. D. Meindl, "A System Level Circuit Model for Multi- and Single-Chip CPU's," *Proceedings of the IEEE International Solid State Circuits Conference*, pp. 308-309, 1987.

[3.90] H. B. Bakoglu, "Circuit and System Performance Limits on ULSI: Interconnections and Packaging," Ph.D. Dissertation, Stanford University, 1987.

[3.91] W. E. Pence, "Electrical, Thermal, and Architecture Aspects of VLSI Packaging and Interconnects for High-Speed Digital Computers," Ph.D. Dissertation, Cornell University, 1989.

[3.92] J. P. Krusius, "System Interconnection of High Density Multi--Chip Modules," *Proceedings of the SPIE Vol. 1390 International Conference on Advances in Interconnects and Packaging*, SPIE vol. 1390, pp. 261-270, 1990.

[3.93] W. P. Birmingham, "Automated Knowledge Acquisition for a Hierarchical Synthesis System", Ph.D. Dissertation, Carnegie Mellon University, 1988. (also published as CMU Research Report No. CMU-CAD-88-25).

[3.94] W. P. Birmingham, A. P. Gupta, and D. P. Siewiorek, "The MICON System for Computer Design," *Proceedings of the 26th Design Automation Conference*, pp 135-140, 1989.

[3.95] W. P. Birmingham, A. P. Gupta, and D. P. Siewiorek, *Automating the Design of Computer Systems - The MICON Project*, Jones and Bartlett Publishers, Boston, MA., 1992

[3.96] T. Sone, K. Kaizu, and M. Hirai, "A Knowledge Base Construction for Technology Planning of Electronic Equipments," *Proceedings of the National Convention of the Institute of Electronics, Information and Communications Engineers (Japan)*, vol. 6, p. 107, Spring 1989.

[3.97] P. A. Sandborn, "A Software Tool for Technology Tradeoff Evaluation in Multichip Packaging," *Proceedings of the Eleventh International Electronics Manufacturing Technology Symposium (IEMT)*, pp. 337-341, 1991.

Chapter 4

Design Partitioning

4.1 Introduction

Complex electronic systems can be implemented in more than one physical unit, i.e., logic and memory can be implemented on one or more chips, chips can be interconnected using one or more modules, and modules can be placed on one or more backplanes which can in turn be placed into one or more enclosures. Throughout the design of a system, engineers are faced with making decisions about how to divide a system's functionality. Partitioning is the design phase in which a designer divides the system's functionality into parts which fit into available physical units. Partitioning is necessary to accommodate varied design constraints, to take advantage of existing components and subsystems, to facilitate parallel development activities, to provide access, to environmentally protect, for testability, for maintainability, and to achieve reconfigurability.

A large number of factors must be considered during partitioning. Some factors are constraints which must be met while others are characteristics which should be optimized. The mechanism which supports the optimization activity is tradeoff analysis (discussed in the last chapter) and the factors which drive partitioning include all of the performance views previously discussed in addition to many more.

Presently, system partitioning is accomplished using highly subjective judgements; a designer divides a system into chips using experience, insight, and luck. For a candidate partitioning, the designer must compute the performance and check to make sure that no design constraints were violated, and the degree to which design target characteristics are minimized or maximized. How close to a minimum or maximum these target characteristics are determines the quality of the partitioning. If the candidate partitioning is inadequate, then a new partitioning must be tried. The process of partition selection and performance evaluation becomes very tedious without tradeoff analysis tools such as those discussed in the last chapter. Even more useful, are algorithms and tools which help designers select the best candidate partitionings thus avoiding an evaluation by trial-and-error.

CAD tools for partitioning at the chip-level have existed for some time. These tools aid designers in determining the best way to divide functionality into chips. Partitioning of logic circuits in order to meet gate and pin count constraints of chips was described many years ago [4.1]. In this approach, a quality function was used to measure the "costs" of partitions, i.e., the compactness and reliability of overall designs. Since that time, tools have appeared which assist designers in partitioning behavioral specifications onto multiple chips while satisfying constraints such as individual chip areas, pin counts, system performance, and system delay, see [4.2] and the references contained therein. At least one tool has been developed to explicitly evaluate the effectiveness of chip partitions. The SPARTA (System PARTitioning Aid) tool developed at GTE does not generate partitionings, it evaluates the user's partitionings in a way similar to the tradeoff analysis tools discussed in the last chapter [4.3]. SPARTA estimates include area, power dissipation, and number of I/O. Other tools already discussed provide the module-level performance information necessary for the designer to choose chip-level partitions but don't necessarily supply a mechanism by which different partitions can easily be described, i.e., AUDiT and MICON.

The next level in the hierarchy is the partitioning of components onto modules or boards. Module partitioning is essentially analogous to its chip-level counterpart although more interdisciplinary constraints must be considered and the basis upon which to partition becomes broader.

4.2 The Basis for Multichip System Partitioning

A complex system can be partitioned in a vast number of ways, for example, ten unique die can be divided amongst two modules or boards in

511 ways[1]. The variety of possible partitions can be categorized into three modes [4.4]:

Functional Partitioning has the goal of decreasing the coupling between substructures so that interface constraints are minimized. This type of partitioning results in a set of substructures optimized to be as stand-alone as possible. The coupling to be minimized may be any application dependent combination of electrical, mechanical, thermal, and informational.

Modular Partitioning has the goal of maximizing the reuse of substructures. This type of partitioning sacrifices optimal separability, capitalizing on previous design experience by repeated application of a single design solution within a system.

Hierarchical Partitioning acknowledges that the creation of a system generally proceeds through several stages. The highest levels consist of relatively stable substructures (stable in a product lifetime sense), while the lower levels consist of substructures which are more rapidly changing. Hierarchical partitioning optimizes the system reconfigurability.

Functional and module partitioning are shown graphically in Figure 4.1. The couplings between the components in Figure 4.1 may be mechanical, chemical, electrical, or thermal. Two components which couple electrically may not couple thermally or mechanically for example, so that a functional partition is view specific, i.e., a functional partitioning based on a thermal view may not represent a functional partitioning from an electrical view.

---

1. The number of combinations is given by,

$$\frac{1}{2}\sum_{r=1}^{n-1}\frac{n!}{n!\,(n-r)\,!}$$

where n is the number of unique die and r is the number of partitions they are divided into assuming that every partition has at least one die in it.

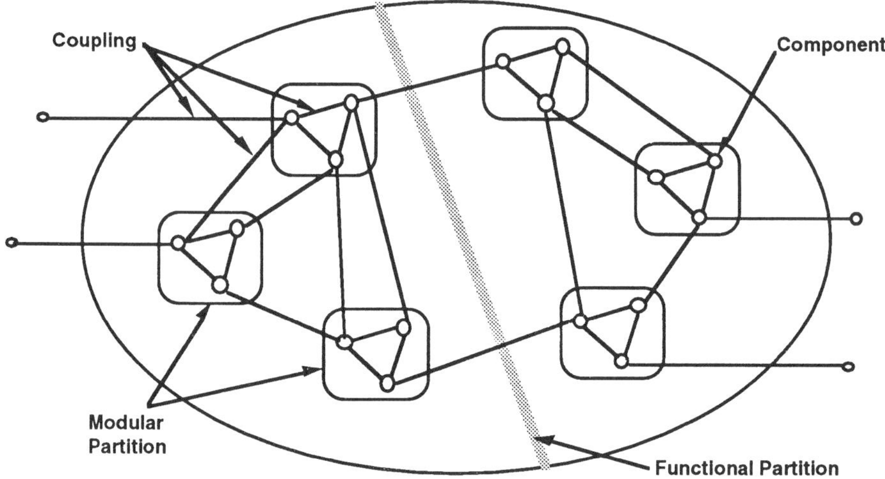

Figure 4.1. Partitioning approaches for multichip systems, [4.4].

4.3 Partitioning Algorithms

In general, formal approaches to partitioning problems represent multivariable optimization problems, and many of the mathematical methods used for their solution are based on more generic optimization approaches which are also used outside the area of microelectronics. The objective of the optimization problem is to place strongly connected components together in such a way as to meet design constraints and targets. Partitioning algorithms are divided into two categories: constructive and iterative improvement methods. A good introduction to basic partitioning methods is contained in [4.5]. In this section, we will only briefly review the best known approaches.

Generic partitioning methods are based on a graph model of the design. The first step is to map a description of the system under design into a graph model. Graph models consist of the following elements:

Physical Components: objects whose describing characteristics are fixed for the purposes of partitioning. In the case of logic partitioning, physical components usually represent flip-flops, registers, adders, etc. In the case of packaging design, physical components may be die over which the designers has no control.

Cutline: division between two physically separate entities.

Edges: physical connections between two components. The edges which cross a cutline represent external connections between two physical entities. For the purposes of graphing, nets with more than two terminals (fanout > 1) are decomposed into several two terminal connections.

Cluster: group of one or more nodes (components).

Objective Function: a function containing weighted terms corresponding to each property to be minimized or maximized in a partitioning activity. The weight indicates the relative importance of one property with respect to another.

Similarly a behavioral description (VHDL for example) can be formulated as a graph partitioning problem. Since packaging and interconnect synthesis usually takes place after chip design and/or selection, we will not discuss behavioral description mapping in this text (see [4.5] for an introductory discussion).

Once a graph is formed, there are two basic partitioning techniques which can be applied. Constructive methods partition the graph by starting with one or more seed nodes and adding nodes one at a time to each partition. Iterative-improvement methods start with some initial partitioning (possibly the result of a constructive method), and successively improve it by moving objects between partitions.

4.3.1 Constructive Methods.

Numerous different constructive methods exist. The simplest method is random selection. This algorithm randomly selects nodes one at a time and places them into clusters of fixed performance (traditionally in chip-level partitioning size is used, however, any other simple metric could be substituted, i.e., cost, weight, exit I/O, mean time to failure, testability fault coverage, etc.) until the target performance is met.

Cluster growth algorithms place objects into a predetermined number of clusters according to some measure ("closeness score") computed from an objective function. The process consists of three activities: 1) seed selection (selection of the seed nodes for each cluster). Seed nodes can be selected randomly, by the designer, or based on a simple set of node attributes. 2) Unplaced node selection determines the order in which to place the remaining nodes into clusters. 3) Node placement adds the unplaced node with the highest closeness score to the appropriate cluster. The process is repeated until all nodes are placed.

Hierarchical clustering considers objects in groups and clusters groups [4.6]. Hierarchical clustering starts by clustering the two closest objects into a group. The group is then clustered with another group or another object on each iteration. The process stops when a single cluster is generated. The design is then partitioned by drawing a cutline through the hierarchical cluster tree. A simple example of hierarchical clustering is contained in [4.5].

Constructive methods are often used to generate initial partitionings for iterative improvement algorithms.

4.3.2 Iterative Improvement Methods

Min-cut partitioning (also known as the Kernighan-Lin algorithm [4.7]) interchanges two groups of nodes between two partitions of equal size to maximize partition performance improvement. The Kernighan-Lin algorithm partitions a graph into two equal subgraphs (an equal number of nodes, m, in each) by minimizing some objective function (traditionally the number of connections between the two subgraphs). In each iteration, the algorithm interchanges n pairs of nodes ($n \leq m$) between the two subgraphs. The process continues until the partitioning is no longer improved. This process is usually performed with multiple sets of initial conditions and the best solution is selected.

Simulated annealing is an optimization mechanism by which to improve the probability of finding the best global partitioning rather than a local optimum. The approaches briefly described above all use simple optimization strategies which can, depending on the complexity of the objective function, easily be drawn into local minima instead of global minima. Simulated annealing is a mathematical technique for solving combinatorial optimization problems [4.8]. The name arises from the close analogy between the mathematical algorithm and the physical processes which take place in a solid material during cooling. Simulated annealing starts with a random initial partitioning and generates a series of moves (moving an object from one partition to another). For each move, a new value of the objective function (called the "cost") is computed and compared against an acceptance criteria. If the value of the objective function is more favorable than the previous value (i.e., the new partitioning is better), the move is accepted. If the value of the objective function is worse than the previous value, the move is accepted or rejected based on a probability of the form,

$$P = e^{-\Delta R/(R_o T)} \tag{4.1}$$

where ΔR is the change in the objective function caused by the move, R_O is the value of the objective function after the previous move, and T is the simulated annealing "temperature." The value of T is scheduled to that after a fixed number of moves its value is reduced by some fraction.

The traditional constraints considered during chip-level partitioning are: chip areas, pin counts, power dissipation (or power dissipation density), and system delays. The performance metrics not treated by traditional chip-level partitioning, but that are necessary for consideration at the module and board level include: cost, testability, reliability, assembly ease, and EMI.

4.3.3 Connection Topology

One partitioning criteria for modules and systems, which is not generally applicable when partitioning functionality into chips, is the connection topological relationship [4.9]. Connection topology is the physical orientation of one partition to another. Three possible topologies that are applicable to inter-module or inter-board connections are: plane-in--plane, edge-to-plane, and edge-to-edge. The connection topology used has no relationship to the number of connections required, but will determine the number of interconnect crossovers. In the present context, a crossover occurs when, in a selected plane, one interconnection between partitions (modules or boards) crosses over another. Crossovers tend to penalize the system size, cost, reliability, and electrical performance. The physical size (and thereby the cost) of a system can grows linearly with the number of connections (N) while the number of possible crossovers grows quadratically,

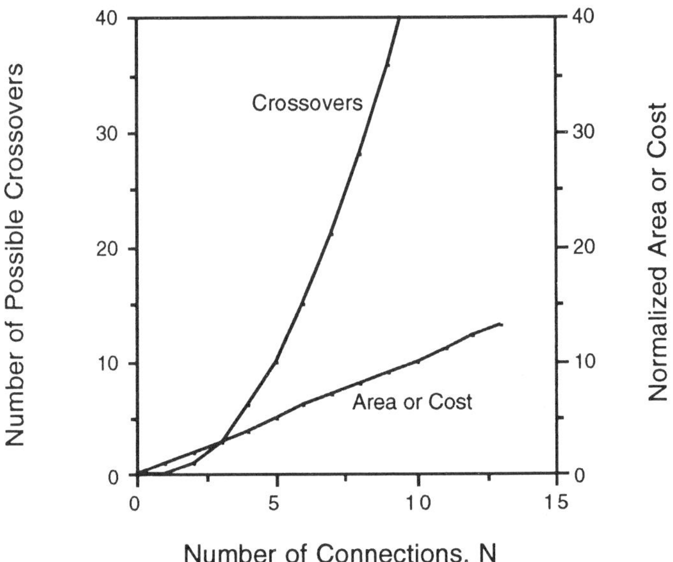

Figure 4.2. Area (cost) and crossover growth with the number of connections.

$$\text{Number of Possible Crossovers} = \frac{N}{2}(N-1) \tag{4.2}$$

Figure 4.2 shows the magnitude of the potential crossover problem.

The number of connection crossovers can be minimized by appropriately partitioning the problem. Parallel bussed connections result in large numbers of crossovers if implemented entirely within a plane.[2] However, if the edge-to-plane topology shown in Figure 4.3 is used, all the crossovers can be eliminated. Edge connector technologies represent folded or double folded edge-to-plane topologies (i.e., a dual-in-line-package, DIP, attached to a board is a folded edge-to-plane topology; a quad package attached to a board is a double folded edge-to-plane topology).

For pseudo-random connection distributions, it is not possible to eliminate the crossovers by selecting the correct topology. The number of crossovers can be minimized, however, using a plane in plane ap

2. For bussed connections every access point in the network is equivalent.

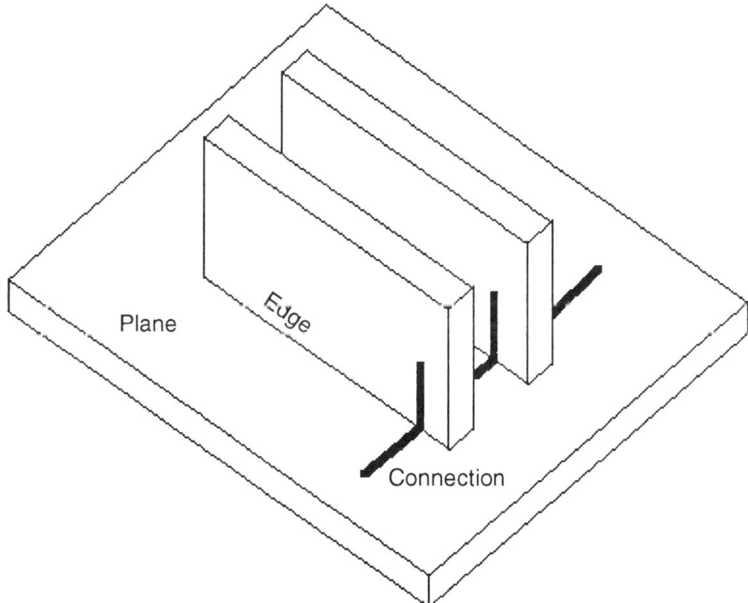

Figure 4.3. Edge-to-plane topology.

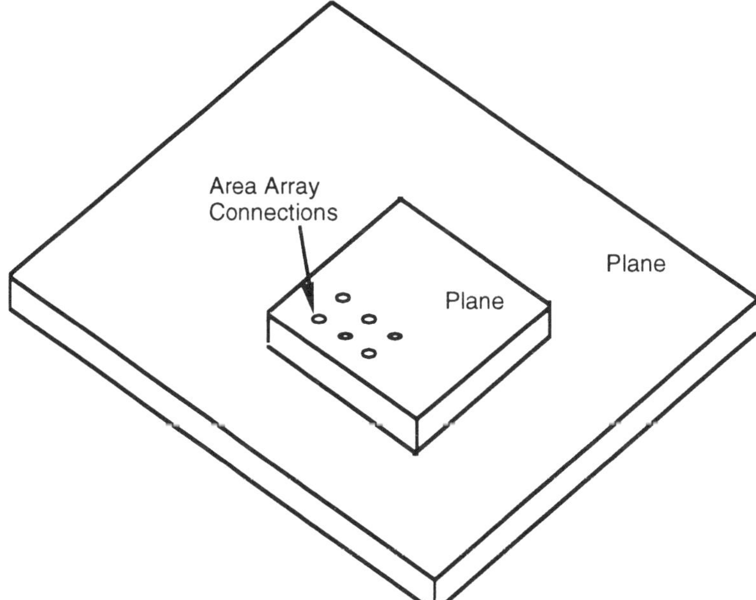

Figure 4.4. Plane-in-plane topology.

proach, Figure 4.4. In general, a uniform area array of connection points on planes minimize the probability of crossovers and perimeter connection points maximizes the crossover probability. Pin grid array (PGA) single chip packages attached to boards represent plane-in-plane topologies.

Edge-to-edge connection topologies allow connections to be permuted. Permuter distribution patterns are widely encountered in telecommunication networks and are can have large numbers of crossovers. The topology permutes because the number of components on one side equals the number of connections per component on the other (see Figure 4.5). Permuted connections occur whenever elements sorted according to one index are redistributed according to a second index.

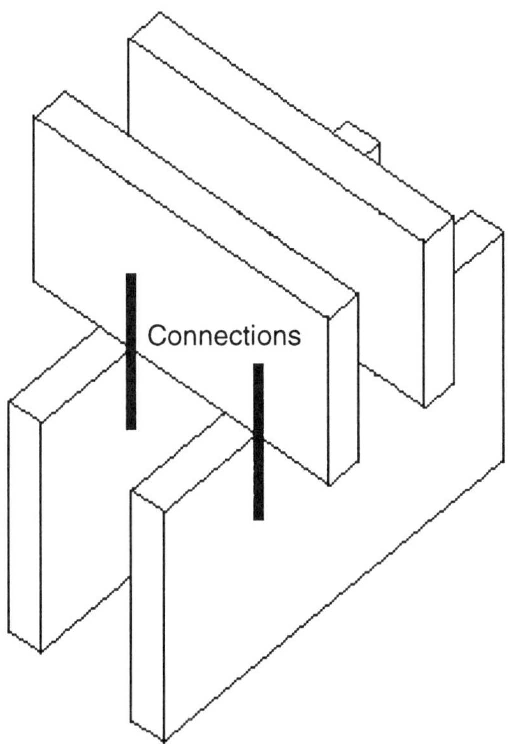

Figure 4.5. Edge-to-edge topology.

4.4 Hardware/Software Partitioning

For today's complex electronic systems, software and hardware design are not independent activities. Wise physical partitioning choices can significantly ease parallel software development requirements. Traditionally, the bulk of computer system functionality is implemented in the software as a sequence of instructions for a general purpose processor. Historically, this has provided the best balance of flexibility, cost, and performance. Semiconductor manufacturers have applied the new economics of VLSI to the production of ever smaller, cheaper, and faster general purpose processors and are relentlessly integrating more and more of the usual computer system hardware elements into ever denser chips. With the arrival of a 64-bit microprocessor generation, it is likely that the continued VLSI focus on general-purpose processing elements will reach the point of diminishing returns, and that further breakthroughs will require an application specific focus.

The most obvious difference between software and hardware solutions to the same problem is the flexibility of the software solution and the inflexibility of the hardware solution. Installation of modified software in the system is accomplished by merely reloading the modified software. Special purpose VLSI requires redesign and refabrication of new hardware and installation of these in the system. While this distinction is becoming less significant (software redesign times may be comparable to VLSI redesign times) still committing a design to hardware freezes many design decisions. Besides flexibility differences, there is a general difference in the view which software and hardware assume. Software design focuses on computation, while hardware design focuses on communication. The limiting factor in software design is usually processor cycles. For hardware, the limitation is communication bandwidth.

The effectiveness of software is measured using algorithm time-complexity (the time needed to perform a task as a function of the problem size) and space-complexity (the amount of memory needed as a function of problem size) [4.10]. Alternatively, hardware algorithms can also be evaluated on the basis of time-complexity, area-complexity, area-time complexity (the product of the area and the time needed as a function of problem size), and area-time squared complexity (the product of the area and the time squared complexity) [4.10]. More generally, volume may be substituted for area. The area or volume bounds are usually memory or I/O count; the area-time bounds are a limit on the total amount of inputs and outputs performed, and the area-time squared bounds are a limit on the communications in the system.

One approach to hardware/software partitioning is shown in Figure 4.6. The system inputs are signal flow graphs and performance specifications. The performance modeler determines the characteristics of the tradeoff space from two functions: decomposition and contention. Decomposition characterizes the extra work (data copying, recomputations) due to partitioning an algorithm. The contention functions characterizes the "cost" of demands for the same resources (memory, buses, or data). The graph partitioner performs partitioning of the signal flow graph using any of a number of techniques mentioned in the last section (i.e., the partitioning activity is identical, the node definition is simply expanded to include software objects in addition to hardware objects). Once a candidate partitioning is identified, hardware and software are synthesized. The performance of the system is evaluated by running the sythesized software on an emulation of the hardware. The system performance is evaluated and feedback is provided to the hardware synthesis system and to the partitioner for design adjustment. The process iterates until an acceptable design is found.

Conventional hardware/software codesign wisdom supports the idea of making the hardware/software partitioning as late as possible in the design process, implying that the system architecture must be made up of stand alone components which can be implemented in either hardware or software [4.11]. In other words, a system's design should be kept as independent of its final hardware/software implementation as long as possible.

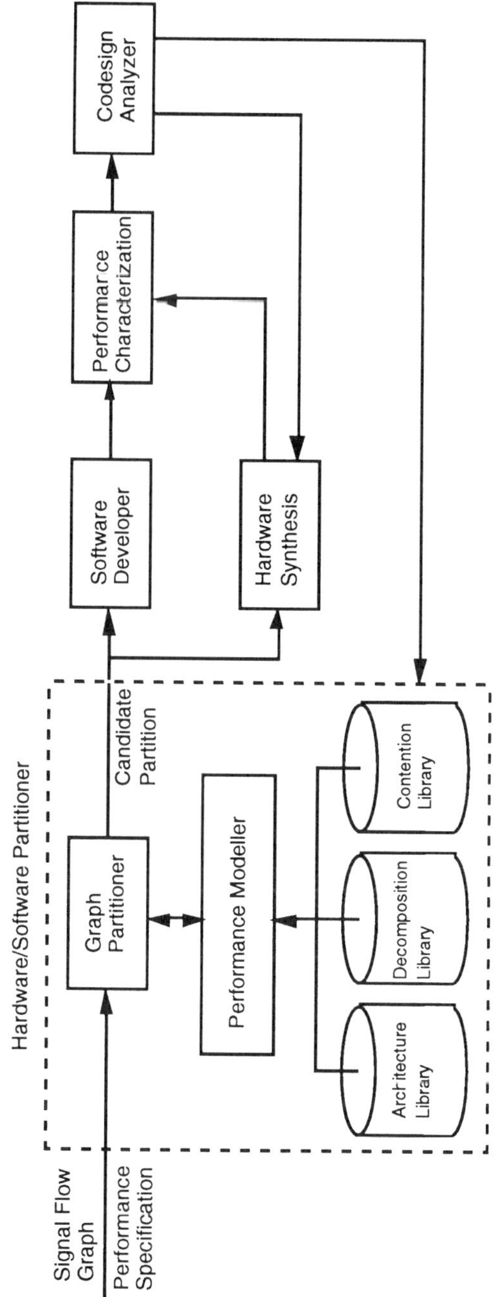

Figure 4.6. Example hardware/software partitioning and analysis system.

4.5 References

[4.1] T. S. Payne and W. M. vanClemput, "Automated Partitioning of Hierarchically Specified Digital Systems," *Proceedings of the 19th Design Automation Conference*, pp. 182-192, 1982.

[4.2] K. Kucukcakar and A. C. Parker, "CHOP: A Constraint-Driven System-Level Partitioner," *Proceedings of the 28th ACM/IEEE Design Automation Conference*, pp. 514-519, 1991.

[4.3] M. L. Resnick, "SPARTA: A System Partitioning Aid," *IEEE Transactions on Computer-Aided Design*, vol. CAD-5, pp. 490-498, October, 1986.

[4.4] D. K. Wilson, "Partitioning-The Payoff Role in Telecommunication Systems Design," *Proceedings of the SPIE International Conference on Advances in Interconnection and Packaging*, SPIE Vol. 1390, pp. 537-547, 1990.

[4.5] D. Gajski, N. Dutt, A. Wu, and S. Lin, *High-Level Sythesis - Introduction to Chip and System Design*, Kluwer Academic Publishers, 1992.

[4.6] S. C. Johnson, "Hierarchical Clustering Schemes," *Psychometrika*, pp. 241-254, September, 1967.

[4.7] K. H. Kernighan and S. Lin, "An Efficient Heuristic Procedure for Partitioning Graph," *Bell Systems Technical Journal*, vol. 49, no. 2, pp. 291-307, February, 1970.

[4.8] S. Kirkpatrick, C. D. Gelatt, and M. P. Vecchi, "Optimization by Simulated Annealing," *Science*, vol. 220, no. 4598, pp. 671-680, 1983.

[4.9] D. K. Wilson, "Topological Aspects of Systems Partitioning," *Proceedings of Design Policy Conference*, Royal College of Art, London, UK, pp. 148-154, 1982.

[4.10] A. C. Hartmann, “Software or Silicon? The Designer's Option,” *Proceedings of the IEEE*, vol. 74, no. 6, pp. 861-874, June, 1986.

[4.11] I. Sommerville, *Software Engineering*, Third Edition, Addison Wesley, 1989.

Chapter 5

Tradeoff Analyses for Multichip Systems

5.1 Introduction

Numerous studies have appeared which compare various packaging technologies [5.1 - 5.16]. These studies contain valuable information about the general applicability of one technology or material over another. Application specific tradeoff studies have also appeared, but are less prevalent than generic treatments [5.17 - 5.23]. These studies present tradeoff analyses for specific real modules.

This chapter presents a set of tradeoff analyses performed using the Multichip Systems Design Advisor (MSDA) tradeoff analysis tool described in Section 3.4.3 in Chapter 3. The first example is a RISC processor module designed and constructed in MCM-L and MCM-D at MCC. The second example details the economic tradeoffs associated with partitioning a system into a few-chip package. The third example compares the tradeoffs between peripheral and area array bonding in an MCM application. The examples do not exercise all the metrology discussed in Chapter 3, but cover a significant cross-section of the primary tradeoff factors.

5.2 RISC Processor Module

The module treated in this example is a RISC processor module with secondary cache designed for the 50-100 MHz range. The module contains 25 chips and up to 48 passive components. The chips consist of 3 VLSI devices (440 I/O and two identical chips with 254 I/O each), 2 buffers, and 20 SRAMs which comprise instruction and data caches. Tradeoff analysis result for a slightly different version of this module appear in [5.17] and [5.21]. The module is an example of a processor module which appears in today's workstation level computers.

Figures 5.1 through 5.5 summarize the tradeoff results for all the cases considered in this example. Because accurate cost information was not available for the chip-first thin film approach, its cost is indicated by a line. All the module costs should be considered on a relative basis and not lent absolute accuracy (i.e., they are appropriate for comparison purposes but do not necessarily represent absolutely accurate magnitudes). Multiple points appear on the graphs for each interconnect technology/assembly approach combination considered. These points represent variations in the design approach such as mixtures of assembly approaches, variations in the number of wiring layers, and variations in the design rules associated with the wiring layers. The following discussion details the variations shown.

Figure 5.1 shows the variation in module cost versus the module area. The printed circuit board (PCB) cases where all die are in single chip packages result in larger systems than multichip module (bare die) approaches due to the large board areas required by the single chip packages. The largest of these solutions has two wiring layers and was interconnect capacity limited, i.e., a maximum of 40% utilization of the wiring resources was assume and all of it was used. All other single chip package solutions had four wiring layers which provided ample wiring resources for this application. In all the cases shown in Figure 5.1, the costs are for first pass assembly and do not include the cost of the bare die.

Four fine line PCB (MCM-L) variations are shown in Figure 5.1. Case 13 has only two wiring layers (150 in/in^2) and is interconnect capacity limited. One of the concerns with this particular module was whether the SRAMs could be acquired in bare die format or not. Cases 14 and 15 represent solutions where the SRAMs were surface mounted in SOJ (Small Outline "J" lead surface mounted single chip package) packages and all other components were wirebonded. The most expensive of the two solutions had four wiring layers and the less expensive only two. In the case of single chip packaged SRAMs, the SOJ packages

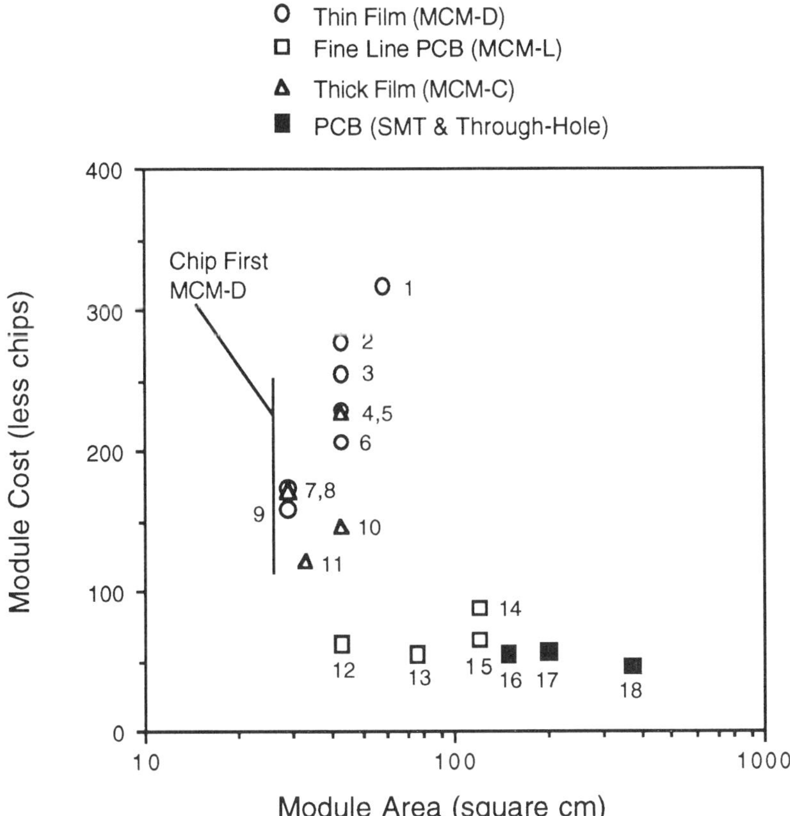

Figure 5.1. Module cost vs. area for an example RISC processor module. The cases shown on this figure and the subsequent figures in this example are:

1) Thin film (MCM-D), 2 wiring layers, 6 mil line pitch on wiring layers (339 in/in^2), all chips TAB bonded except for buffers (wirebonded), uniform thermal vias throughout interconnect.

2) Thin film (MCM-D), 2 wiring layers, 3 mil line pitch on wiring layers (677 in/in^2), all chips TAB bonded except for buffers (wirebonded).

3) Thin film (MCM-D), 2 wiring layers, 6 mil line pitch on wiring layers (339 in/in^2), all chips TAB bonded except for buffers (wirebonded).

4) Thin film (MCM-D), 2 wiring layers, 3 mil line pitch on wiring layers (677 in/in^2), all chips wirebonded.

5) Thick film (MCM-C), 6 wiring layers (343 in/in^2), all chips wirebonded.

6) Thin film (MCM-D), 2 wiring layers, 6 mil line pitch on wiring layers (339 in/in^2),

7) Thin film (MCM-D), 2 wiring layers, 3 mil line pitch on wiring layers (677 in/in^2), all chips C4 (flip chip) bonded.

8) Thick film (MCM-C), 6 wiring layers (343 in/in^2), all chips C4 (flip chip) bonded.

9) Thin film (MCM-D), 2 wiring layers, 6 mil line pitch on wiring layers (339 in/in^2), all chips C4 (flip chip) bonded.

10) Thick film (MCM-C), 4 wiring layers (229 in/in^2), all chips wirebonded.

11) Thick film (MCM-C), 4 wiring layers (229 in/in^2), all chips C4 (flip chip) bonded.

12) Fine Line PCB (MCM-L), 4 wiring layers (300 in/in^2), all chips wirebonded.

13) Fine Line PCB (MCM-L), 2 wiring layers (150 in/in^2), all chips wirebonded.

14) Fine Line PCB (MCM-L), 4 wiring layers (300 in/in^2), all chips wirebonded except SRAMs (SOJ single chip packages).

15) Fine Line PCB (MCM-L), 2 wiring layers (150 in/in^2), all chips wirebonded except SRAMs (SOJ single chip packages).

16) PCB (SMT & Through-Hole), 4 wiring layers (133 in/in^2), components assembled on both sides of the board, VLSI chips in PGAs, buffers in QFPs, and SRAMs in SOJs.

17) PCB (SMT & Through-Hole), 4 wiring layers (133 in/in^2), components assembled on one side of the board, VLSI chips in PGAs, buffers in QFPs, and SRAMs in SOJs.

18) PCB (SMT & Through-Hole), 2 wiring layers (68 in/in^2), components assembled on one side of the board, VLSI chips in PGAs, buffers in QFPs, and SRAMs in SOJs.

provide enough fanout that two wiring layers provides sufficient interconnect resources to wire the problem.

Four thick film solutions are shown in Figure 5.1. The two 42 cm^2 solutions are assembled using wirebonding and the two smaller solutions use C4 (flip chip) bonding. When wirebonding is used, the module size is the same whether four or six wiring layers are used. When C4 is used, the four wiring layer solution is interconnect capacity limited and therefore slightly larger than the six wiring layer solution.

There several different thin film results in Figure 5.1. In all the thin film cases, enough wiring was present in two wiring layers to route the module (i.e., thin-film solutions were never wiring limited for this application). Chip-first MCM-D results in the smallest module due to the small spacing between the die (20 mil minimum was allowed). The chip-first results are plotted as a line because the costs were not computed. C4 (flip chip) bonded MCM-D solutions are nearly as small (29 cm^2). The highest cost C4 version used MCM-D with a 3 mil line pitch on the wiring layers and the less expensive version used a 6 mil line pitch on the signal routing layers. In general, the chip-last thin film results indicate that there is no need to use 3 mil (pitch) design rules for this application (i.e., 6 mil (pitch) rules provide more than enough wiring at less cost). In case 1, the thermal vias blocked enough wiring tracks that the module size became interconnect capacity limited.

From Figure 5.1 it can be seen that the best cost/size tradeoff for this application may be wirebonding on fine line printed circuit board (case 12). If size is the most critical driver, then chip-first thin film is a good choice.

Figure 5.2 shows a summary of the address line delay versus module area for a subset of the solutions presented in Figure 5.1. The delay in

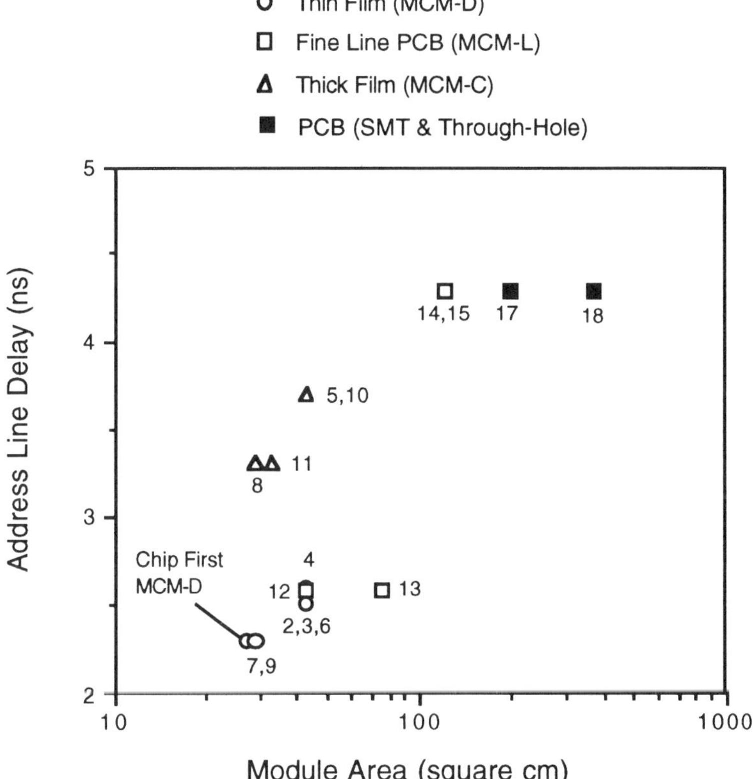

Figure 5.2. Module area vs. packaging delay for an example RISC processor module. See Figure 5.1 for an explanation of the cases plotted.

this case corresponds time-of-flight, RC charging, and includes effects of impedance mismatch and reflections (no crosstalk or switching noise effects are included). The critical path considered is a driver on one of the VLSI chips driving an address line fanned out to twelve of the SRAMs (4 pF each). The delays quoted in this section are for the "furthest" (in a delay sense) SRAM. Figure 5.2 shows the large delay penalty for using single chip packaged SRAM (cases 14, 15, 17, and 18). Not only does the use of single chip packages increase the size of the module and cause net lengths to increase, but the single chip packages add capacitance, which increases RC charging delays.

If enough wiring resources are available so that the size of the module is not interconnect capacity limited, then the PCB (MCM-L) implementations can accommodate delays just as small as the chip-last thin-film implementations if similar dielectrics and assembly approaches are used. However, if the SRAMs have to be placed in single chip packages for surface mounting, a delay penalty is assessed due to the larger module size and increased capacitive load on the address lines.

Thin film solutions which use flip chip bonding methods have the smallest delays due to smaller module sizes and slightly smaller capacitive loads. The chip-first thin-film implementation is the fastest because of its extremely small size. In general, the 3 mil line pitch implementations are slightly slower than the 6 mil implementation due to the larger resistive loss in their signal lines (greater RC charging delay). The thick film modules are slower because of the longer time-of-flight due to the larger dielectric constant of the ceramic material (ε_r = 9.5).

Figure 5.3 shows an example electrical analysis output from the MCC MSDA tradeoff analysis tool. The top plot shows the driver waveform and the resulting signal at two of the receivers on the critical net. The bottom plots are Bode magnitude and phase relations at the receivers.

Thermal analysis of this example vehicle includes the estimation of internal and external thermal resistances and the comparison of those resistances to heat removal capabilities of various heat sink technologies. The first step in the analysis is the creation of plot shown in Figure 5.4. Generating this relation depends on a component's power dissipation, size, and thermal budget (allowed temperature rise over ambient, 55 degrees C for all chips in this example). The packaging technology used

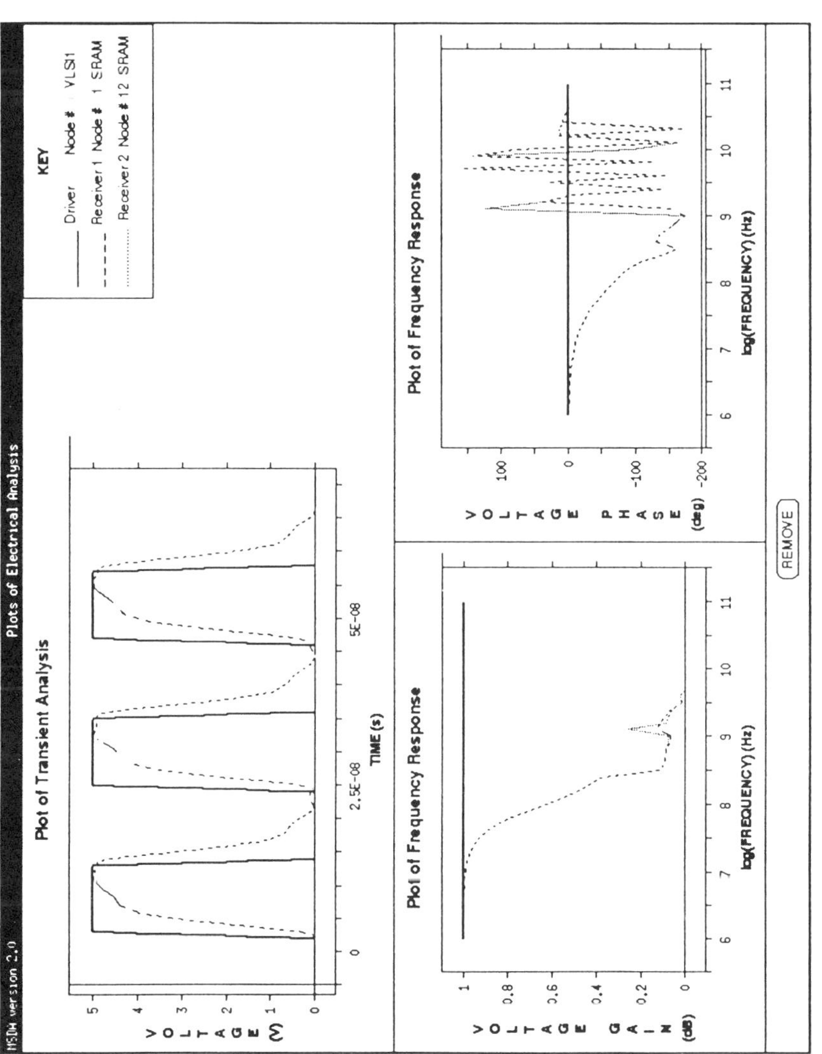

Figure 5.3. Example electrical analysis output from the MCC MSDA tradeoff analysis tool.

for the application will determine the operating point on the horizontal axis for each component and the corresponding applicable cooling approaches (see Section 3.3.3 for an explanation of how to interpret Figure 5.4).

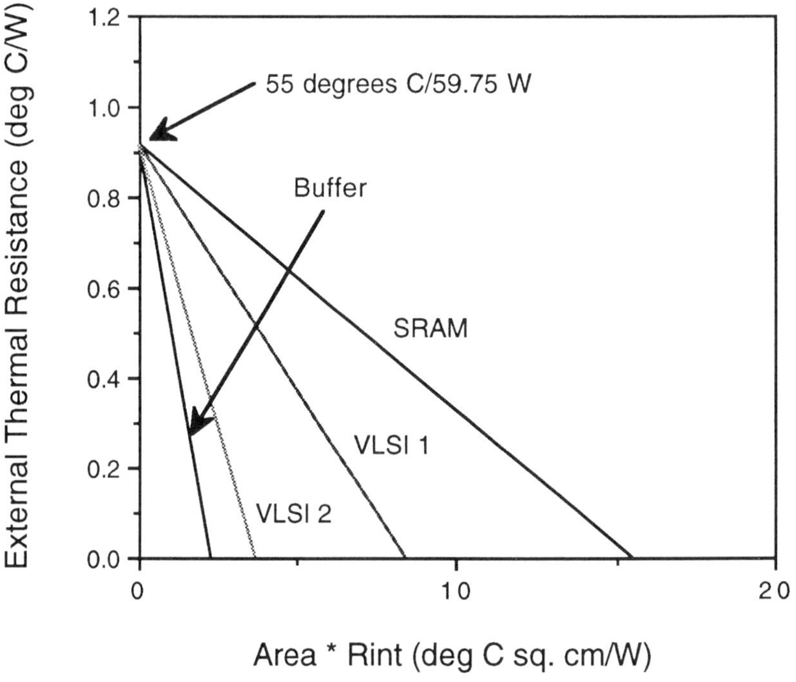

Figure 5.4. The theoretical cooling limits for each of the components in the example module. The SRAMs dissipate 0.75 W each and have a die area of 0.21 cm^2; the VLSI 1 chip dissipates 13 W and has a die area of 1.98 cm^2; the VLSI 2 chips dissipate 15 W each and have a die area of 1 cm^2; the buffer chips dissipate 0.875 W each and have a die area of 0.37 cm^2.

Figure 5.5 shows a summary of the thermal operating points for the buffer chips in this design. Only the buffer results are shown because they were thermally the most problematic due to their high power dissipation density. The only cooling path considered in these results is through the substrate under the die.

The chip-first thin film implementation has slightly less internal thermal resistance than the other implementations since the path through the "substrate" is through the component, its attach, and the machined carrier it is glued into, and does not include the interconnection layers. The rest of the results appear in two clusters; all the cases in the upper cluster include thermal vias uniformly distributed under the buffer die and, in the face-up bonding cases, a heat spreader the size of the die. The lower cluster of cases have no thermal vias. Cases which fall below zero

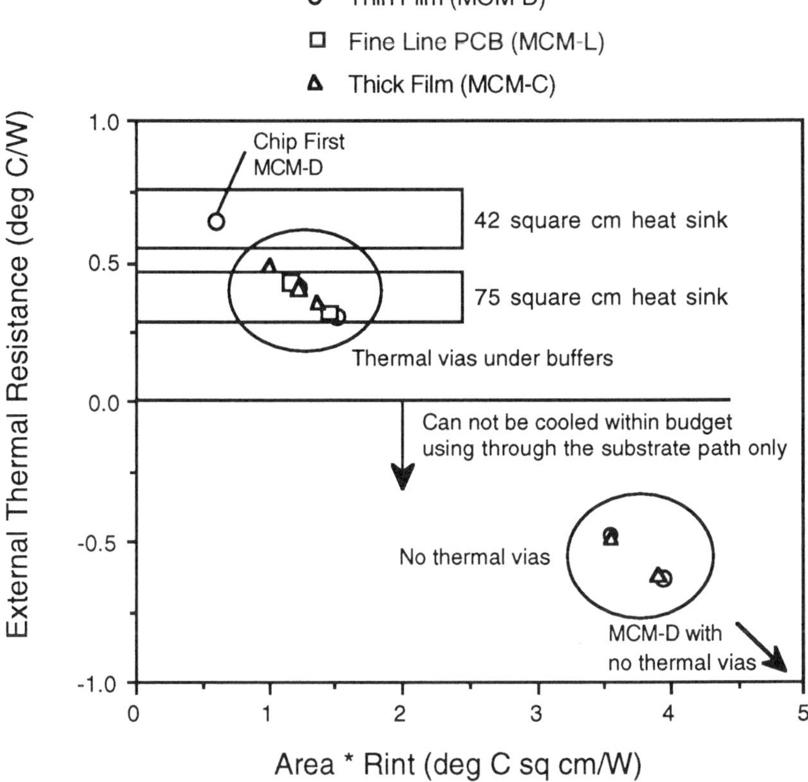

Figure 5.5. The thermal performance through the substrate under the buffer chips in the RISC processor module. A thermal budget limiting the maximum temperature rise over ambient of 55 degree C was assumed. The internal thermal resistances are computed through a heat sink attach layer. Only bare die cases are shown (single chip package cases are not included). Only through substrate cooling options are compared in this figure.

external thermal resistance can not be cooled to the budget using the through substrate path only. The performance of two extrusions are shown. In both cases, the air velocity was varied from 5 to 20 m/s and the extrusions were 750 mils high, 5 mils thick, had a 100 mil center to center pitch, and were made from aluminum.

Figure 5.6 shows a photograph of the RISC processor module considered in this example. The module shown was constructed on a thin film interconnect, the SRAMs and VLSI devices are TAB bonded and the buffer chips (top of the photograph, to the right and left of the large chip) are wirebonded.

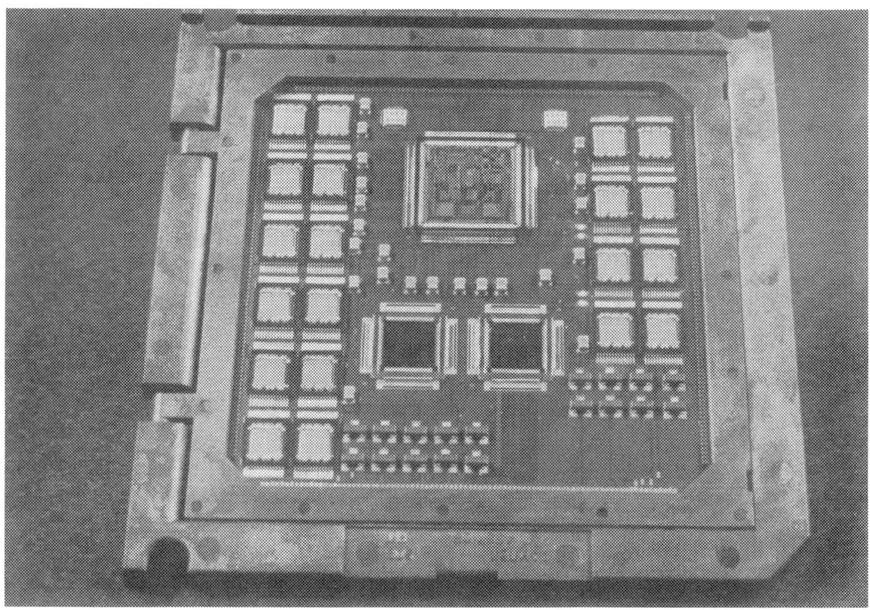

Figure 5.6. Thin film multichip module implementation of the RISC processor module treated in this section. All chips are TAB bonded except two buffers which are wirebonded.

Sometimes it is tempting to perform tradeoff analyses by considering only a portion of the module, leaving out details which would appear to contribute second order effects such as neglecting passive components which are present on every module. Figure 5.7 replots Figure 5.1, where each variation includes two data points. The closed points represent the module without any passive components (resistors, capacitors, diodes, or test pads) included. The open points represent the complete module. Obviously, the addition of the passive components increases the size and cost of the module in all cases. The objective of this figure is to show that the cost of the high-density interconnect approaches increases much more quickly than that of more traditional approaches when low I/O count components are added. From Figure 5.7 we see that without the passive components there are thin film cases which appear

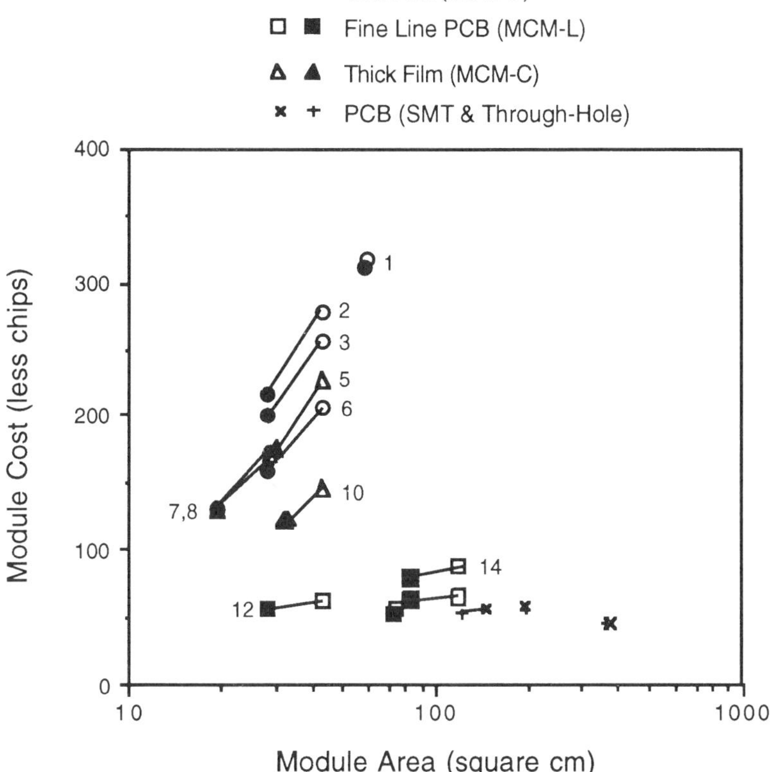

Figure 5.7. Module cost vs. area for an example RISC processor module with second level cache. Each variation shown on the figure includes two data points. The closed points represent the module without any passive components (resistors, capacitors, diodes, or test pads) included. The open points represent the complete module. See the caption of Figure 5.1 for identification of the cases plotted.

to be good cost/size tradeoffs (at least as good as the best thick film solutions and smaller). Once passive components are added, two of the MCM-C (thick film) solutions cost less, and one is almost as small as the best chip-last thin film solution. These results illustrate that in many applications, the presence of the passive components can not be ignored in the technology selection process. These results also suggest that the economical use of high density interconnection approaches for some applications requires the development of techniques to embed passive components in the interconnect and/or the placement of passive components on active die surfaces.

5.3 Few-Chip Packaging

As the number of I/O of chips going into conventional single chip packages, i.e., quad flat packages (QFPs) and Pin Grid Arrays (PGAs), have increased, the package size has grown extensively. In a commercial 304 lead QFP, the ratio of the silicon area to the package area is 12 mm square/40 mm square, or 9%. The majority of the real estate used in a single chip package is for fanout and escape channels to get to a pitch that can be handled with existing pick and place and bonding equipment. Therefore, filling existing single chip packages with multiple chips begins to become cost and size effective when the silicon density in the package can be increased above that of single chip packages. Besides reducing system size and cost, few-chip packages allow improvements in electrical performance by grouping performance sensitive portions of a system into MCMs while not penalizing system assembly. Few-chip packages also offer the potential for improved testability over solutions which do not use single chip packages. Section 1.1.1 in Chapter 1 includes a brief discussion of few-chip packages.

In this example, we consider an 80486 based PC design with a cache controller. A block diagram of the system is shown in Figure 5.8. The board considered contains a 80486 microprocessor, numeric coprocessor, a data cache consisting of five 8k x 8 SRAMs, a three chip cache controller set, and a two chip ROM. The DRAM block is a separate SIMM card, which can be connected outside of the board being studied here.

The objective of this example is to determine the tradeoffs associated with packaging two or more of the system's components in a few-chip

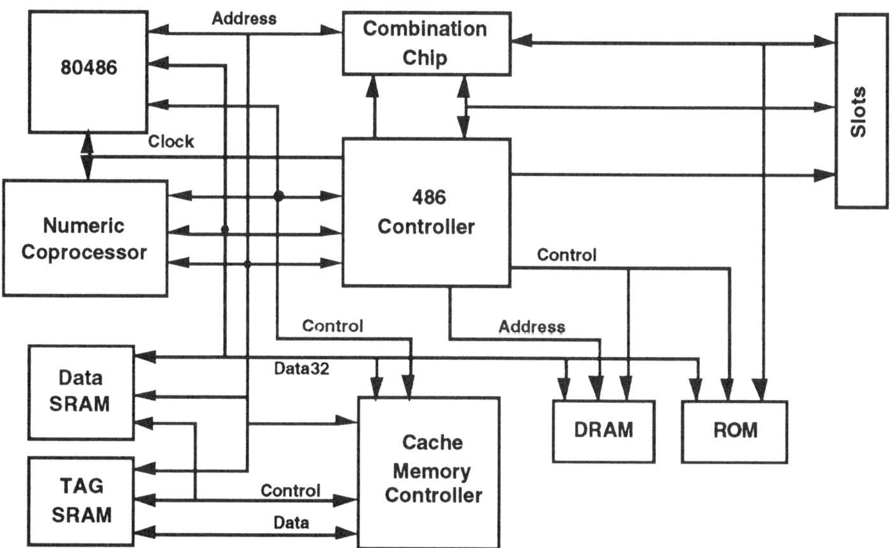

Figure 5.8. System diagram for a 80486 based PC motherboard with a cache controller. The entire board (excluding the DRAM) consists of approximately 250 nets and 170 off-board signal I/O.

package. For this application, performance targets are met using traditional packaging approaches (i.e., all chips in single chip packages) and thermal challenges are confined to the microprocessor and coprocessor chips. Therefore, the quantitative results concentrate on cost versus size tradeoffs. This situation is obviously not shared by every application; those which must be packaged in MCMs for electrical performance reasons may not allow their critical components to be partitioned amongst different packages.

The characteristics of each of the chips along with the block diagram shown in Figure 5.8 were the analysis inputs. The baseline case for comparison consisted of all the chips packaged in their own single chip packages (the 80486 and coprocessor in ceramic PGAs and all other chips in surface mount packages) mounted on an FR-4 printed circuit board. Various permutations of placing two or more of the system components on an MCM in a few-chip package were explored. Note that not all possible permutations were considered - we did not allow the two ROM chips or the five SRAMs to be split up, and the 80486 and coprocessor were never included in few-chip packages due to their higher power dissipations. Table 5.1 summarizes the characteristics of the few-chip modules. Only wirebonding on fine line printed circuit boards was considered inside the few-chip packages due to an emphasis on low cost and quick insertion (migration) into high volume manufacturing quantities. The fine line board had 2 mil lines and spaces and a varying number of layers (between 2 and 6 layers were used for routing signals in the different cases shown in Table 5.1).

The simple cost model used to generate the results in Table 5.1 assumed steps for dispensing die attach material, die placement, post cure, plasma clean, and wirebonding. The packaged QFP costs were generated by summing the cost of the completed module, the cost of an appropriate leadframe as well as leadframe attach, and the cost of the molding and materials for the package. The leadframe costs assumed are approximately half those for similar I/O counts used in single chip packages because few-chip package boards require less material and can be stamped since they do not require as fine a lead pitch. All results assumed volumes in excess of 100,000 and domestic labor rates.

The results in Table 5.1 have limited usefulness by themselves. In order to assess the size/cost applicability of each partition, the whole system cost and size must be considered. Table 5.2 and Figure 5.9 show the system size/cost tradeoff for the cases defined in Table 5.1.

The results in Table 5.2 and Figure 5.9 indicate that there are solutions containing few-chip packages which are less expensive than the traditional assembly approach with all chips in their own packages. Depending on the constraints imposed by the designer, few-chip packages may or may not be a viable alternative for this system.

The least expensive cases which include a few-chip package are those in which the few-chip package contains the greatest number of chips. The smallest systems also corresponded to the cases in which the few-chip package contained a large number of chips. Adding system components to the few-chip package is only of benefit as long as the module in the few-chip package remains small enough to be housed inside a package of reasonable size. The maximum QFP size we have allowed in this study is 44 x 44 mm.

Table 5.1. Summary of the few-chip packages considered in this study. The module size and cost represent the properties of the unpackaged MCM which interconnects the selected components. The QFP properties are for few-chip packages, with the MCM inside of them (less the chip cost). The number of I/O listed refers to the number of signal I/O leaving the few-chip package. The total number of I/O leaving the package was taken to be 1.2 times the number of signal and control I/O. The last column on the table gives the cumulative cost of the single chip packages (less the chip costs) used in the traditional packaging approach for the components in the selected few-chip package.

Contents of the Module (Few-Chip Package)	Number of Signal Routing Layers in the Board	Module Size (mm)	Module Cost, less chips ($)	QFP Size (mm)	QFP Packaged Cost, less chips ($)	Each chip in its own single chip package, less chips ($)
Combination Chip, 486 Controller, Cache Memory Controller, Data SRAM, TAG SRAM (235 I/O, 254 nets)	6 4	36x36 42x42	17.50 13.96	40x40 44x44	19.25 15.83	7.76
486 Controller, Cache Memory Controller (219 I/O, 219 nets)	4	27x29	6.98	36x36	8.48	5.63
Combination Chip, 486 Controller (235 I/O, 243 nets)	4	24x24	5.95	28x28	7.05	5.63
Memory Controller, Data SRAM, TAG SRAM (96 I/O, 107 nets)	4 2	28x22 43x43	6.58 6.08	32x32 44x44	7.88 7.95	2.13
Combination Chip, Cache Memory Controller, Data SRAM, TAG SRAM (158 I/O, 169 nets)	4	35x23	8.13	40x40	9.88	4.01

Contents of the Module (Few-Chip Package)	Number of Signal Routing Layers in the Board	Module Size (mm)	Module Cost, less chips ($)	QFP Size (mm)	QFP Packaged Cost, less chips ($)	Each chip in its own single chip package, less chips ($)
486 Controller, Cache Memory Controller, Data SRAM, TAG SRAM (208 I/O, 229 nets)	4	36x36	11.21	40x40	12.96	5.88
Fabricated Example. Combination Chip, 486 Controller, Cache Memory Controller (246 I/O, 254 nets)	4	36x36	9.95	40x40	11.70	7.51
Combination Chip, 486 Controller, Cache Memory Controller, ROM (246 I/O, 254 nets)	6 4	36x36 39x39	16.45 11.93	40x40 44x44	18.20 13.80	7.67
486 Controller, ROM (208 I/O, 208 nets)	4	26x26	6.60	28x28	7.70	3.91
Combination Chip, 486 Controller, ROM (235 I/O, 243 nets)	4	36x36	9.92	40x40	11.67	5.79
486 Controller, Cache Memory Controller, Data SRAM, TAG SRAM, ROM (228 I/O, 239 nets)	4 6	42x42 36x36	14.01 17.61	44x44 40x40	15.88 19.36	5.88

Table 5.2 - Summary of the system sizes and costs versus few-chip packages chosen for this study. The first case represents the system with no few-chip package present, i.e., all chips in single chip packages. The costs include the board, single chip packages, and the few-chip package modules described by Table 5.1.

Contents of the Few Chip Package	System Size (sq. inches)	QFP System Cost, less chips ($)
All components in their own single chip package	28.06	48.57
Combination Chip, 486 Controller, Cache Memory Controller, Data SRAM, TAG SRAM (235 I/O, 254 nets)	15.72 16.31	50.2046.21
486 Controller, Cache Memory Controller (219 I/O, 219 nets)	25.77	50.55
Combination Chip, 486 Controller (235 I/O, 243 nets)	28.46	49.93
Memory Controller, Data SRAM, TAG SRAM (96 I/O, 107 nets)	35.17 35.17	51.88 52.37
Combination Chip, Cache Memory Controller, Data SRAM, TAG SRAM (158 I/O, 169 nets)	28.76	48.99
486 Controller, Cache Memory Controller, Data SRAM, TAG SRAM (208 I/O, 229 nets)	19.35	47.71
Fabrication Example. Combination Chip, 486 Controller, Cache Memory Controller (246 I/O, 254 nets)	20.86	48.85
Combination Chip, 486 Controller, Cache Memory Controller, ROM (246 I/O, 254 nets)	15.98 16.01	52.47 47.73
486 Controller, ROM (208 I/O, 208 nets)	22.14	51.66
Combination Chip, 486 Controller, ROM (235 I/O, 243 nets)	17.24	50.74
486 Controller, Cache Memory Controller, Data SRAM, TAG SRAM, ROM (228 I/O, 239 nets)	16.29 16.29	47.83 51.53
Combination Chip, ROM (168 I/O, 168 nets)	34.05	53.24

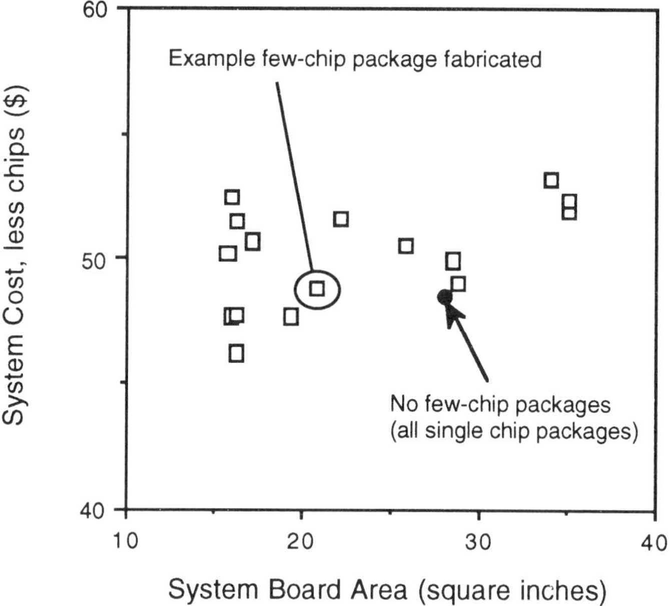

Figure 5.9. The system cost/size tradeoff of the various partitionings described in Table 5.2.

Cost and size obviously do not represent all the concerns which must be taken into account in deciding on feasible partitions for a system. Many systems have thermal constraints or electrical performance requirements which may rule out many (or all) potential partitionings. In this study, we avoided thermal problems by leaving the "hot" chips out of all potential partitions and electrical performance targets were already met by the lowest performance case (all chips in their own single chip packages). The concerns we have not addressed quantitatively are testability, reliability, manufacturability, and business aspects. In reality, the final partitioning of the system shown in Figure 5.8 was chosen for business reasons based on a *Modular Partitioning* constraint (as discussed in Section 4.2 in Chapter 4). In order to maximize the reuse of substructures, the Combination chip, 486 Controller and Cache Memory Controller were combined in a few-chip package (this is the seventh case in Table 5.1). The manufacturer of the cache controller chip set may have wanted to let the system designer have flexibility to choose the size of the data and DRAM caches, and add their own ROM. The following discussion details the implementation of this few-chip package.

The example few-chip package marked on Figure 5.9 has been constructed. The module is based on a chip set from VLSI Technology, Inc. An Intel 486 CPU controller chip set was designed and integrated into a three chip module. The chip set consists of a 40 MHz 486 system controller chip with 208 I/O that generates 1 W; a cache controller chip with 128 I/O that generates less than 1 W, and combination chip with 100 I/O which generates approximately 0.5 W. All chip designs are based on 1 μm CMOS design rules and the on- and off-module frequency is 40 MHz with signal pulses having 1 ns edges. There are six decoupling capacitors that are also used inside the module for regulating switching noise.

The module requires an interconnect technology which can route 240 nets on a 36 x 36 mm board. A four layer fine line laminated board was designed which has 2 mil lines and spaces with 10 mil through vias on 18 mil pads. All chip dimensions and interconnect design rules are identical to those considered in the tradeoff study presented in the last section. There is a slight variation between the number of nets assumed in the theoretical evaluation and the actual number in this example due to variations in the number of passive components included. Typical of many few-chip applications designed by the authors, most of the routing is done on the top layer, though some nets were routed in a stripline fashion in the inner layers. The top side metallization consists of a chip-pad, wirebonding pads, and 304 10 mil wide peripheral pads on 0.45 mm pitch. An array of thermal vias are placed under each die pad.

The substrate is housed in a standard 40x40 mm QFP that has 304 leads. A Cu-alloy etched leadframe is soldered to the board using a high temperature solder material. After conventional die attach and gold wirebonding, the module is tested and molded. The part is fully encapsulated using standard low stress molding compounds. The external package I/O are on a 0.5 mm pitch.

A photograph of the completed multichip module and the packaged system is shown in Figure 5.10. The design is laid out to accommodate up to 6 discrete capacitors though they may not all be necessary. The 4-layer board consists of a power and ground plane, though some minimal routing is done on these layers as well. This particular partitioning

Figure 5.10. Photograph of the few-chip package containing a three chip cache controller set. The lower left shows the fine line PCB interconnect prior to bonding the components. The right side shows a completed MCM attached to a leadframe, and the upper left shows the final molded package. (Courtesy of the MCC Single/Few-Chip Packaging Project)

of the overall design shown in Figure 5.8 is of interest because it allows the system designer the flexibility to decide the data cache and DRAM memory sizes while conveniently grouping the cache controller functionality.

A cost effective market pull for MCM technology requires finding a way to introduce MCMs into the mainstream of electronic packaging without substantial changes in the ways vendors, systems houses, and suppliers do business; this opportunity may develop through the use of few-chip packages. Quad flat packages and pin grid arrays are good examples of a packaging technologies that are well understood, and pervasively used, which can be modified to house more than one chip. Ideally, the few-chip package would cost less than its components packaged alone, but most of the savings are realized when it is included in a system. The system lifetime cost which acknowledges the effects of test, repair, upgradability, system modularity, and maintainability improve with the use of few-chip packaging.

5.4 Peripheral vs. Area Array Bonding

The partitioning of functionality into chips is a fundamental design concern for high performance systems. The challenge is to determine the optimal number of chips for a given functionality, based on a cost/performance tradeoff. Putting a large amount of functionality onto a single chip may provide electrical performance and system size advantages, but often results in large die with low yields and high costs. Alternatively, realizing the same functionality using a large number of small die, results in less costly die, at the expense of a larger system size and possible performance degradation. In addition to cost and performance issues, the testability of a single large die is often more complex than that of several small die that perform the same function.

The optimal integration density (number of logic gates or memory bits per chip) depends on a large number of tightly coupled design goals, including: cycle time, noise, power dissipation, cost (yield), physical size, and testability. Mainframe computers have traditionally been built using a large number of bipolar chips containing a relatively small number of gates that consume a large amount of power. Personal computers are built using highly integrated (usually single) CPU systems with much smaller power dissipations, lower costs, and poorer performance than high-end systems. These market segments are making functional partitioning tradeoffs.

In light of these tradeoffs, the objective of this example is to determine the optimal number of chips necessary for partitioning a given functionality into, as a function of peripheral and area array die formats. Direct chip attach methods used in MCMs include peripheral bonding approaches such as wirebonding and tape automated bonding (TAB), and area array approaches such as flip chip or array TAB (ATAB). Peripheral approaches require all of the die I/O to be in a single (or at most double) row around the perimeter of the die. Area array approaches distribute the die I/O over the entire face of the die. In general, flip chip approaches result in smaller systems (since die can be placed closer together) and higher performance systems (since the effective inductance associated with the chip bonds is reduced and line drivers can potentially be smaller). On the other hand, area array bonding requires extensive die preparation (today, this is only economical before wafers are diced) and the infrastructure to support peripheral bonding is considerably greater than that for area array (i.e., very few IC manufacturers are producing area array format die).

In order to assess the full impact of peripheral versus area array bonding on a system, a concurrent analysis of system size, performance, and cost including chip cost, test, and rework must be performed. Analyzing system costs while neglecting the chips will not generally lead to a correct relative cost comparison between these packaging alternatives. Similarly, analyzing chip costs to meet a specified functionality without considering the cost of packaging and the costs associated with test and rework can be equally misleading; i.e., a selected functional partitioning may appear economical at the chip level but could result in a more expensive system once test and rework costs are accounted for.

The analysis approach employed for this study makes use of a combination of existing software modeling tools and algorithms. Figure 5.11 shows the process used to determine the overall module cost for various partitionings and bonding approaches. The die characteristics were predicted using the SUSPENS model [5.24]. Our implementation of SUSPENS accepts the total number of gates and chips into which to equally divide the gates as inputs (the gates are assumed to be indistinguishable). The model computes the number of signal and control I/O from Rent's rule and the die core size using an estimation of the average wire length which assumes a die core size that is interconnect capacity limited. The number of power and ground I/O that a chip requires is application specific; however, for this study we have assumed constant values for the signal I/O to ground I/O ratio: 4.2 for peripheral bonding cases (similar to the DEC Alpha chip, [5.25]), and 6.0 for area array cases (an equal number of power and ground I/O are assumed). A larger signal to ground ratio is allowed for area array bonding due to the lower effective inductance associated with flip chip bonding (due to shorter bond lengths). With the die core size and the total number of I/O determined, the final die size can be computed. The size estimations for peripheral and area array die are similar to the formulations developed in

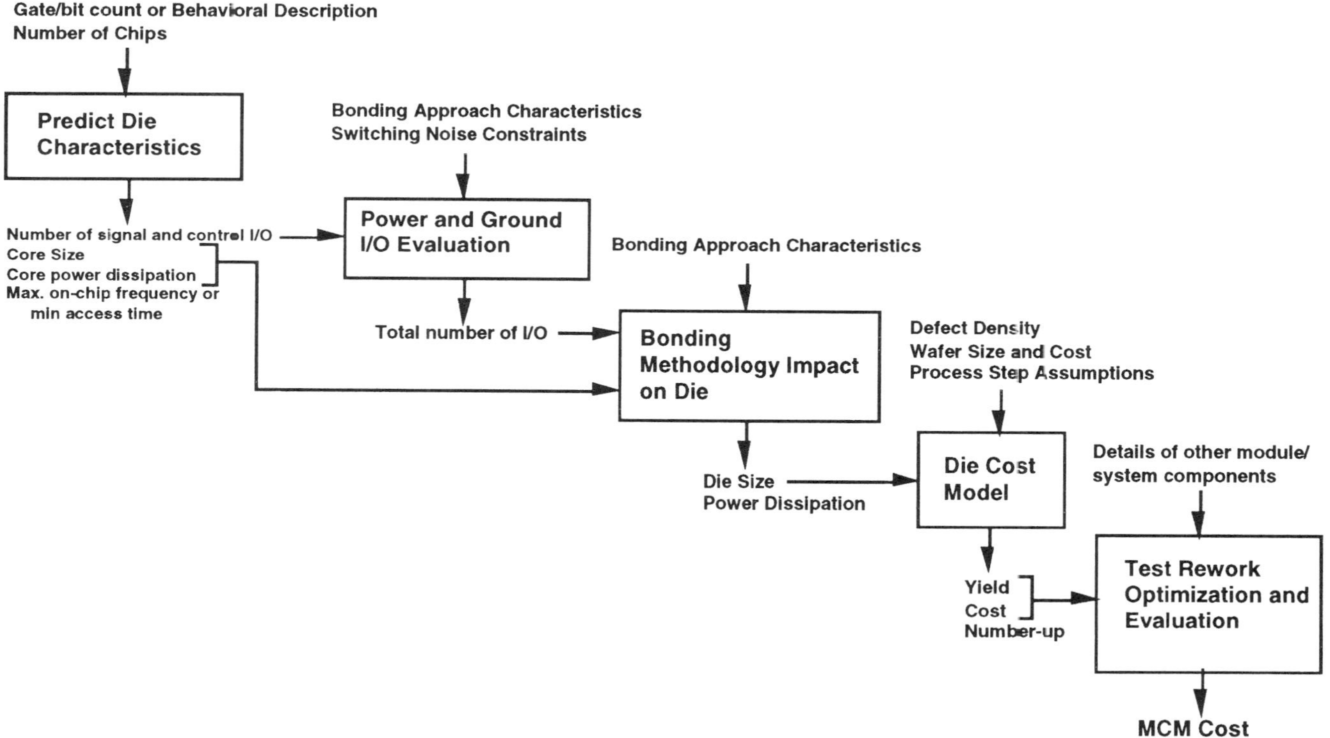

Figure 5.11. The methodology used to determine the MCM cost for various partitioning and bonding approaches.

[5.26]. For a peripherally bonded die, the area is given by the maximum of two limitations:

$$\text{peripheral area 1} = (2\text{length}_{pad} + \text{pitch}_{pad}\lfloor N_c/4 \rfloor)^2 \quad (5.1a)$$

$$\text{peripheral area 2} = N_c \text{width}_{pad} \text{length}_{pad} + (1 + \kappa N_c)\,\text{area}_{core} \quad (5.1b)$$

where

length_{pad} = length of a peripheral bond pad on the die
width_{pad} = width of a peripheral bond pad on the die
pitch_{pad} = minimum center-to-center pitch of peripheral bond pads on the die
N_c = total number of die I/O
κ = the fraction increase in the core die area necessary to accommodate redistribution of one I/O to the periphery of the die
area_{core} = core die area.

Equation (5.1a) is the I/O limited chip area and (5.1b) is the peripheral redistribution limited area, both assume a single row of bond pads. For an area array bonded die, the die area is given by the maximum of two limitations:

$$\text{area array 1} = (\text{pitch}_{pad}\lceil \sqrt{N_c} \rceil)^2 \quad (5.2a)$$

$$\text{area array 2} = N_c \text{width}_{pad} \text{length}_{pad} + \text{area}_{core} \quad (5.2b)$$

where

length_{pad} = length of an area array bond pad on the die
width_{pad} = width of an area array bond pad on the die
pitch_{pad} = minimum center-to-center pitch of area array bond pads on the die.

Equation (5.2a) is the I/O limited chip area and (5.2b) is the bond pad area limitation, assuming that active circuitry cannot be placed under the bond pads. Subtracting (5.2b) from (5.1b) yields the size gradient results derived in [5.26], with the exception of the "outer core" redistribution term, which is accounted for in our derivation by the (5.1a) limitation.

After the final die size is determined, the cost of the die is computed. The yield of die on the wafer was computed using Murphy's yield law with a fixed defect density. The number-up on the wafer was computed assuming a fixed minimum spacing between die (50 mils). In order to compute the die yield and cost at the beginning of the assembly process, the methodology shown in Figure 3.39 was followed. The die yield after wafer test is computed from the test coverage (fraction of the defects identified in the test) and the actual yield of the die on the wafer. After sawing, the known defective die are scrapped and the rest are sent on to burn-in.

The characteristics of the rest of the module (size, type of substrate, wiring capacity, etc.) are estimated using the tradeoff analysis tool described in Section 3.4.3. Included in this analysis is a detailed estimation of module costs, along with the cost of assembly, test, and repair as discussed in [5.27].

Table 5.3. Characteristics of the test module used for the example in this section. The yield corresponds to the yield of the die at the start of assembly (i.e., at the end of the process shown in Figure 3.39). The third chip (CPU) is the functionality which is the subject of the partitioning in this example, the numbers in parenthesis indicate its characteristics in single chip form, where p = peripheral and aa = area array. The critical assumptions used to generate the results in this example are included in Table 5.4.

Chip	Quantity	Signal and Control I/O	Total I/O	Die Size (mils)	Power Dissipation (W)	Yield (fraction)	Cost ($)
SRAM	10	62	100	470 x 300	1.5	0.9972	29.63
ASIC	1	484	750	600 x 600	50	0.9967	80.77
CPU	Variable (1)	(527)	(781 p) (701 aa)	(790 x 790 p) (744 x 744 aa)	(58.4)	(0.9968 p) (0.9970 aa)	(169 p) (176 aa)

Table 5.4. Critical analysis assumptions.

Property	Value
Bond Pad Size (peripheral)	2.5 x 3 mils
Bond Pad Size (area array)	5 x 5 mils
Min. Bond Pad Pitch (peripheral)	4 mils
Min. Bond Pad Pitch (area array)	10 mils
Wafer Defect Density	3 defects/square inch
Processed Wafer Cost	$800
Wafer Diameter	6 inches
Unusable Wafer Boarder	0.4 inches
Minimum Space Between Die on Wafer	50 mils
Wafer Bumping Cost (per wafer)	$200
Defects Added by Wafer Bumping	0.2 defects/square inch
Wafer Test Cost	$3.00 per die
Wafer Test Coverage	80%
Die Burn-in Cost	$2.60 per die
Induced Failures at Burn-in	10%
Die Test Cost	$3.00 per die
Die Test Coverage	99%
Substrate Yield	99.999%
Assembly Yield	99.995%
Repair Yield	80%
Module Test Cost	$75.00 per module
Module Test Coverage	95%
Die Logic Type	CMOS
Min. Feature Size on Die	0.7 μm
Quantity	50,000 modules

Results for a conventional chip in which there is negligible unused die area are presented in the following paragraphs. In all cases, the chip (or gate set) that was studied was assumed to be part of a processor module that included an additional large chip (an ASIC of some type) and ten SRAMs. The characteristics of the chips in the module are given in Table 5.3.

To begin the analysis, we fix the number of signal I/O of the CPU chip to 527 and adjust Rent's rule so that there is no empty area inside the peripherally bonded chip. The number of gates in the chip in this case is approximately 2,388,000 (assuming 0.7 μm design rules). The results for this case are then generated by fixing the number of gates at the above value and varying the number of CPU chips used to realize them.

The results of this analysis are shown in Figures 5.12 through 5.19.

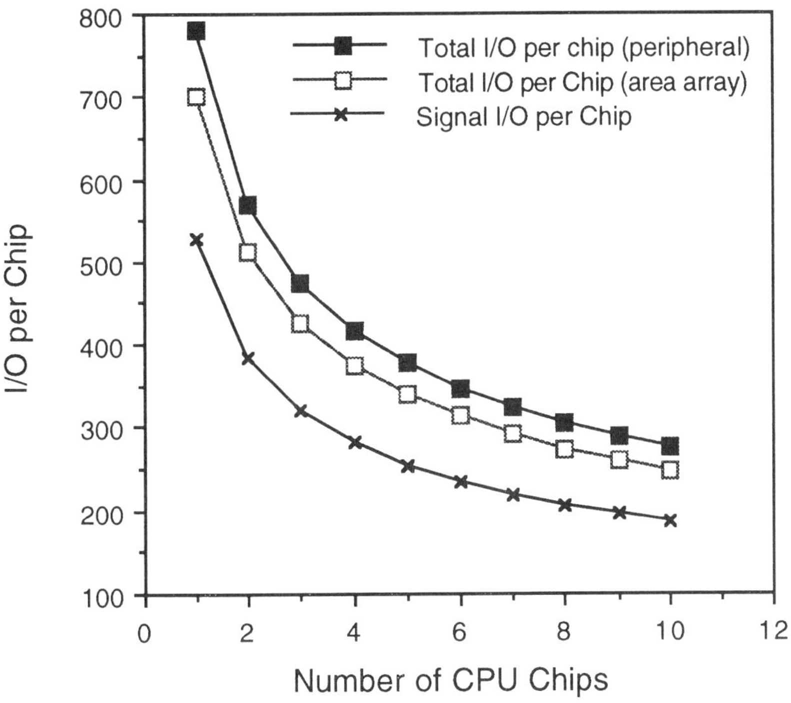

Figure 5.12. I/O per CPU chip as a function of the number of chips the CPU is divided into. Signal I/O refers to all I/O which are not powers or grounds.

Figure 5.12 shows how the number of signal and total I/O per chip varies as the CPU functionality is divided into various numbers of die. The number of signal I/O is the same in the peripheral and area array cases and the number of total I/O in the area array case is less than the peripheral case (discussed above). Figure 5.13 shows the power dissipation per chip (Per Die) and the total power dissipation of all the chips used to realize the CPU (Die Set). A chip's power dissipation is proportional to its number of gates, so power dissipation drops as the number of gates drops. The power dissipation per gate, however, increases as the size of the die decreases, since the average dynamic power dissipation is proportional to on-chip frequency and the maximum on-chip frequency is increasing due to smaller die areas. The result is a net increase in the

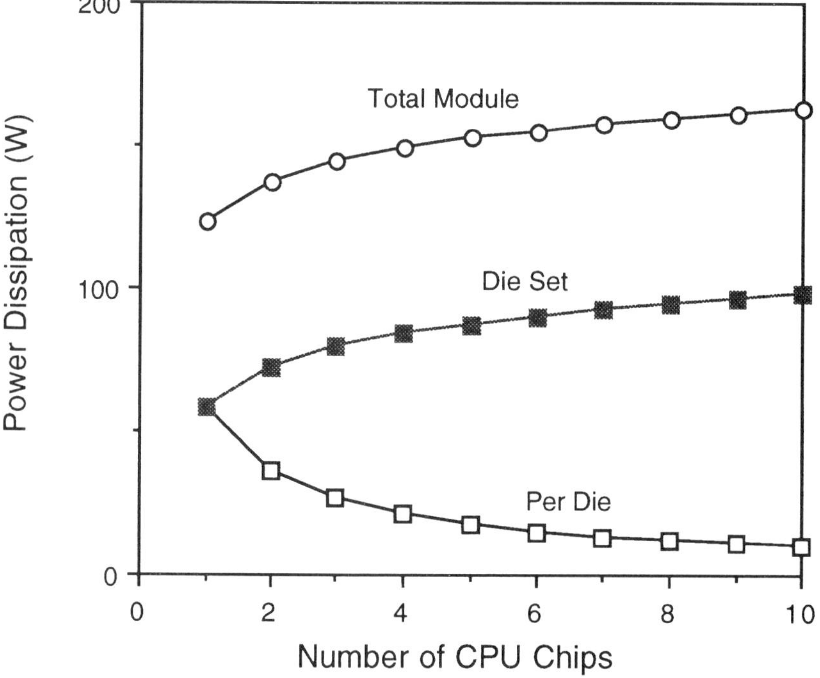

Figure 5.13. Power dissipation per CPU chip and total power dissipation for the set of chips the CPU is divided into (the power dissipations of all other components in the MCM are included in the Total Module curve).

power dissipation of the system as the number of chips increases (this argument assumes that we will want to run the chips faster).

Figure 5.14 shows the die area variation. Both the peripheral format chip and the area array format chip track the core area. The area array chip is smaller than the peripheral chip due to the extra area required for the redistribution of connections from inside the core of the die to its periphery. As the I/O per area decreases, the percentage of the die area necessary for redistribution decreases and the peripheral and area array format chips limit to the same size.

Figures 5.15 and 5.16 summarize the cost results. Figure 5.15 shows that the difference in cost per die for peripheral and area array format die is minimal. When only one die is used for the CPU, the peripheral die cost is actually less than the area array die cost because the cost of bumping the wafer containing the area array die is amortized over relatively

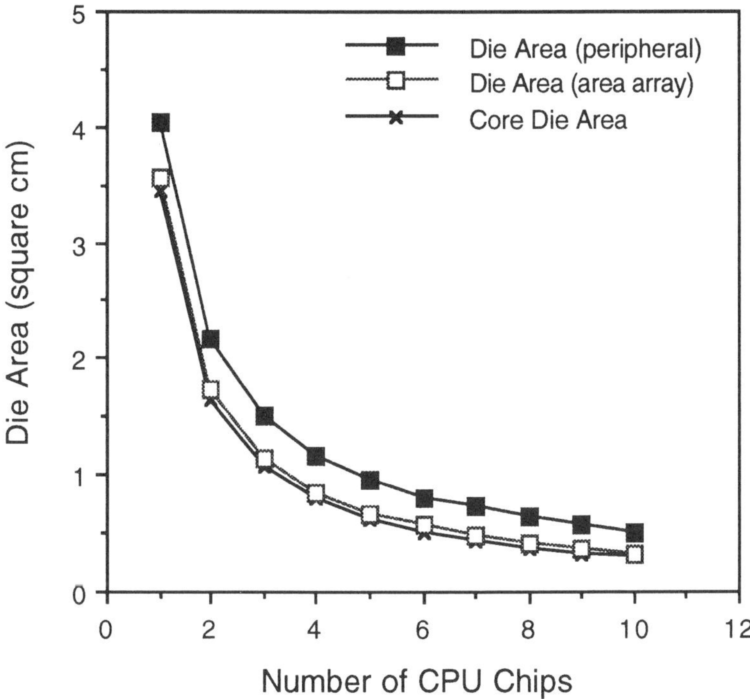

Figure 5.14. Single CPU die area and core die area as a function of the number of chips the CPU is divided into.

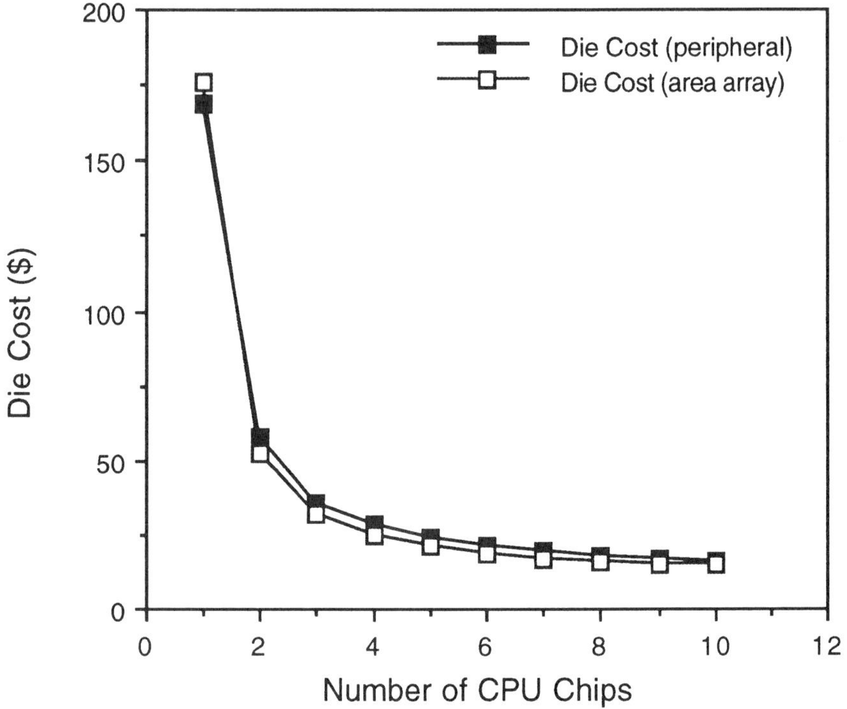

Figure 5.15. Single CPU die cost as a function of the number of chips the CPU is divided into.

few working die per wafer (due to larger die size and therefore fewer die/wafer). As the number of working die per wafer increases, the cost savings for having slightly smaller area array die than peripheral die, more than offsets the bumping cost incurred by each die. For this example, our primary interest is not the cost of a single die but the cost of the completed multichip module when assembly, test, and repair effects are considered.

Figure 5.16 shows the MCM cost as a function of the number of chips the CPU functionality is divided into. In all cases, an 11 layer (3 mil lines and spaces) low temperature cofired ceramic (LTCC) interconnect was assumed. For this application, the cost of the chips is dominant (\$473 - \$550), the cost of the LTCC interconnect is approximately \$45 in all cases, and the rest of the MCM cost in Figure 5.16 is assembly, repair, and testing. The figure shows that for this application, the opti-

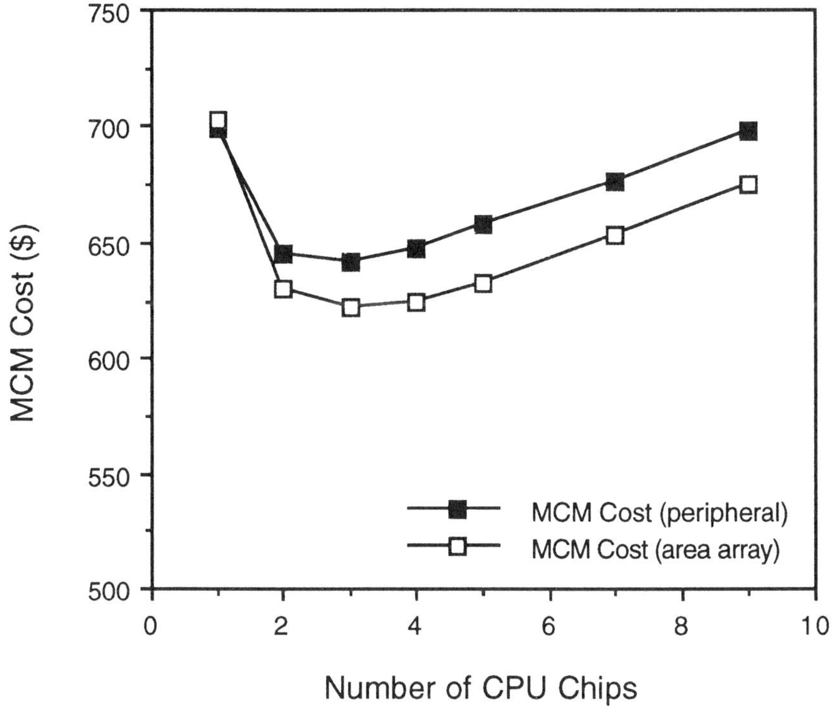

Figure 5.16. Completed MCM cost. The MCM cost includes test and repair and the costs associated with all components (see Table 5.3).

mal number of chips into which to divide the CPU functionality, from a cost standpoint, is three. With the exception of using one chip, the area array approach results in lower MCM costs for all partitionings than the peripheral case. The results in Figure 5.16 are application specific. An application where the die do not contribute as great a fraction of the total module cost could display different characteristics.

Cost may be the greatest driving factor in many MCM designs, but it cannot be considered in isolation. We must also consider the size, electrical, and thermal performance of the resulting MCM. Figure 5.17 shows the overall MCM size as a function of the number of chips the fixed functionality is divided into. There are several competing effects which determine overall module size. As the number of chips increase, their individual size decreases. However, the results in Figure 5.17 become "saw tooth" shaped in the peripheral bonding case due to how the

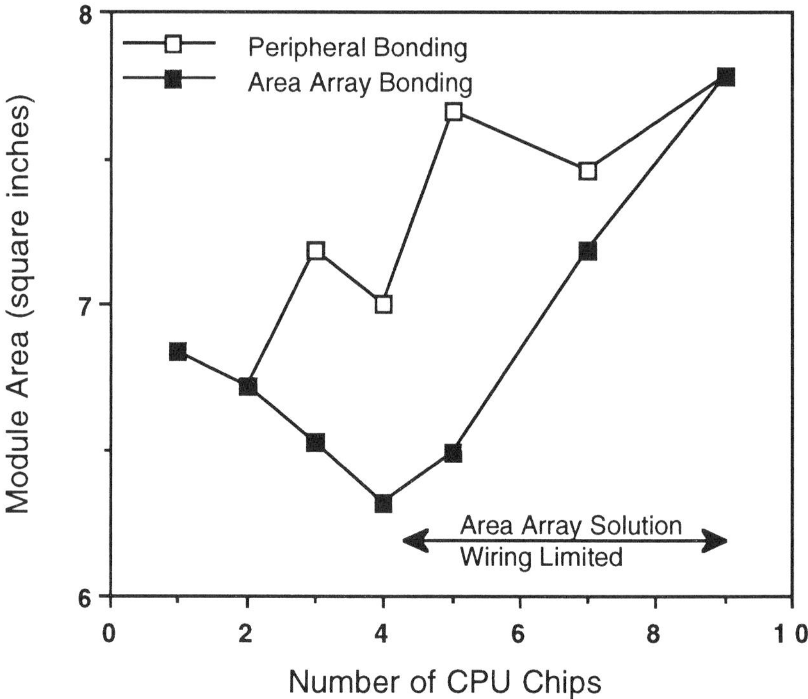

Figure 5.17. Overall MCM area as a function of the number of chips the CPU is divided into.

chips tile together. For example, three chips may require more space than four slightly smaller chips because of how they fit together. In this study, we required the CPU chips to fit together so that the corner-to--corner Manhattan distance associated with the CPU chips was minimized. In the area array case, the module size decreases until the module becomes limited by the amount of interconnect wiring available.

The area array bonding case generally results in a smaller MCM, however, if the module size is interconnect capacity limited, the size advantages of area array bonding over peripheral bonding may be negated. It is also evident that the number of components the functionality is divided into can significantly impact the module size. In this application an even number of components is preferable. However, another application, could prefer an odd number depending on the size and number of other components in the module.

Figure 5.18 shows an electrical performance figure of merit for the MCM. The packaging delay includes the effects of RC charging, time-of-flight, and reflections. The delay was computed for a length equal to the furthest corner-to-corner separation between chips that contain the

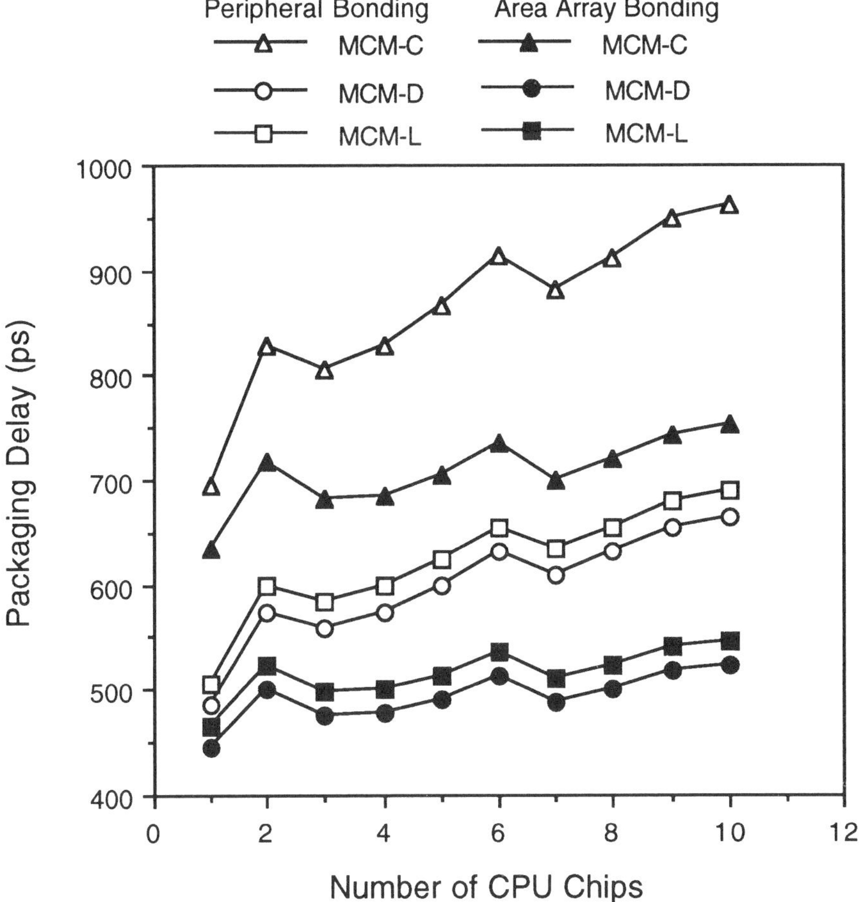

Figure 5.18. Electrical performance as a function of the number of chips the CPU is divided into, and MCM interconnect technologies. The characteristics of the interconnects are: MCM-L - FR-4 board, $\varepsilon_{eff} = 4.5$, signal line width = 8 mils, thickness = 1.4 mils; MCM-D - $\varepsilon_{eff} = 3.4$, signal line width = 15 µm, thickness = 5 µm; MCM-C - $\varepsilon_{eff} = 9.5$, signal line width = 4 mils, thickness = 0.7 mils.

CPU functionality (the path does not correspond to a real electrical connection). The path has only one load on the end (3 pF). A driver with $R_{on} = 50\ \Omega$ was assumed in all cases and all signal lines had a 50 Ω characteristic impedance. A minimum spacing between components of 100 mils in area array cases and 150 mils in peripheral bonding cases was assumed.

The peripheral bonded case always result in more delay than the area array bonded case due to the longer path length. The curves are not smooth because of the tiling issues as discussed previously. The path lengths for the different interconnect technologies were assumed to be the same. The delay differences are due to differing dielectric constants.

Figure 5.19 shows thermal results. Two thermal paths to heat sinks were considered: through the interconnect under the die and above the die. The minimum allowable external thermal resistances plotted in Figure 5.19 were computed using (3.40).

On Figure 5.13, the total module power dissipation was 65 W plus the power dissipation per CPU die multiplied by the number of chips used to realize the CPU. A thermal budget of 55°C maximum rise over ambient was allowed in all cases. A 25 mil thick silicon die was assumed in all cases. In the face-down above substrate cases, 3 mils of thermal grease (1.1 W/mK) was assumed between the die and the heat sink. The face-up above die cases required thicker thermal grease to accommodate the extra space needed for bonding (i.e., wire bond loops, etc.), making them impractical.

The thermal performance of the peripheral bonding cases is better than the area array bonding performance. This is because of the larger die size (resulting in less internal thermal resistance). Note that the above die peripheral cases plotted are face-down (not face-up). The peripheral through interconnect cooling solution included a heat spreader the size of the die and 4 mil diameter thermal vias on a 12 mil pitch. For this particular application, the through substrate and above substrate paths for the peripherally bonded die result in approximately the same performance. In the area array bonding case, the above substrate path is preferable. Also note that the solutions which contain fewer chips are easier to cool (but run at a slower frequency).

Several quantities used in this example have been treated as constants, when in reality, they are a function of the die size and its I/O

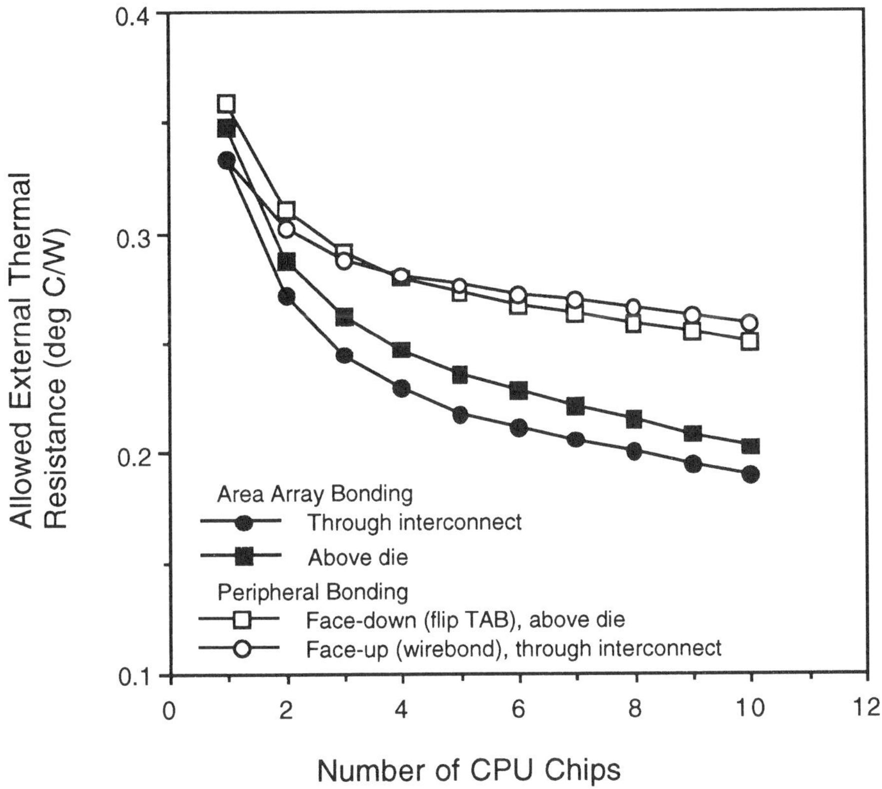

Figure 5.19. Thermal performance as a function of various cooling paths. All results are for an 11 layer LTCC interconnect. The higher the allowed external thermal resistance the better.

count. These quantities include the wafer test cost (assumed to be a function of the number of I/O), wafer test coverage, die test cost, assembly yield, and repair yield. Extensions to this work should concentrate on formulating functional relations for these quantities in order to obtain more accurate comparison results.

5.5 References

[5.1] D. Balderes and M. White, "Packaging Effects on CPU Performance of Large Commercial Processors," *Proceedings of the 35th Elec-*

tronic Components Conference, pp. 351-356, 1985.

[5.2] H. B. Bakoglu and J. D. Meindl, "A System Level Circuit Model for Multi- and Single-Chip CPU's," *Proceedings of the IEEE International Solid State Circuits Conference*, pp. 308-309, 1987.

[5.3] C. A. Neugebauer and R. O. Carlson, "Comparison of Wafer Scale Integration with VLSI Packaging Approaches," *IEEE Transactions on Components, Hybrids, and Manufacturing Technology*, vol. CHMT-10, no. 2, pp. 184-189, June, 1987.

[5.4] J. P. Krusius and W. E. Pence, "Analysis of Materials and Structure Tradeoffs in Thin and Thick Film Multi-Chip Packages," *Proceedings of the Electronic Components Conference*, pp. 641-646, 1989.

[5.5] V. K. Nagesh, D. Miller, and L. Moresco, "A Comparative Study of Interconnect Technologies," *Proceedings of the International Electronic Packaging Symposium (IEPS)*, pp. 433-443, 1989.

[5.6] C. A. Neugebauer, "Materials for High-Density Electronic Packaging and Interconnections in the Higher Packaging Levels," *J. of Electronic Materials*, vol. 18, no. 2, part 2, pp. 229-239, March, 1989.

[5.7] R. Kaw, "Comparison of Chip Crossing Delay in Various Packaging Environments," *Proceedings of the International Conference on Computer Design*, pp. 233-236, 1989.

[5.8] L. L. Moresco, "Electronic System Packaging: The Search for Manufacturing the Optimum in a Sea of Constraints," *IEEE Transactions on Components, Hybrids, and Manufacturing Technology*, vol. 13, pp. 494-508, September, 1990.

[5.9] R. Hannemann, "Interconnects and Packaging for Highly Integrated Systems," *Presented at SPIE International Conference on Advances in Interconnect and Packaging*, 1990.

[5.10] J. P. Krusius, "System Interconnection of High Density Multi-- Chip Modules," *Proceedings of the SPIE International Conference on*

Advances in Interconnects and Packaging, vol. 1390, pp. 261-270, 1990.

[5.11] L. L. Moresco, "System Interconnect Issues for Sub-Nanosecond Signal Transmission," *Proceedings of the SPIE International Conference on Advances in Interconnects and Packaging,* vol. 1390, pp. 202-213, 1990.

[5.12] C. A. Neugebauer, R. A. Fillion, W. Daum, and M. Gdula, "The Single Chip Versus Multichip Packaging Option for Digital CMOS in the 1990's," *IEEE Transactions on Components, Hybrids, and Manufacturing Technology*, Vol. 15, no. 5, pp. 915-921, October, 1992.

[5.13] G. Messner and W. Smit, "Equations for Selection of Cost-Efficient Interconnection Designs," *Proceedings of Electronic Component and Technology Conference*, pp. 10-16, 1992.

[5.14] L. W. Schaper, "Meeting System Requirements Through Technology Tradeoffs in Multi-Chip Modules," *Proceedings of the International Electronic Packaging Symposium (IEPS)*, pp. 25-33, 1990.

[5.15] M. Terasawa and S. Minami, "A Comparison of Thin Film, Thick Film, and Co-Fired High Density Ceramic Multilayer with the Combined Technology: T&T HDCM (Thin Film and Thick Film High Density Ceramic Module)," *International Journal for Hybrid Microelectronics*, vol.6, no.1, pp. 607-615, October, 1983.

[5.16] A. Iqbal, M. Swaminathan, M. Nealon, and A. Omer, "Design Tradeoffs Among MCM-C, MCM-D and MCM-D/C Technologies," *Proceedings of the IEEE Multi-Chip Module Conference (MCMC)*, pp. 12-17, 1993.

[5.17] P. A. Sandborn, "A Software Tool for Technology Tradeoff Evaluation in Multichip Packaging," *Proceedings of the International Electronics Manufacturing Technology Symposium*, pp. 337-341, 1991.

[5.18] M. M. Salatino and R. C. Braken, "Die and MCM Test Strategy: The Key to MCM Manufacturability," *Proceedings of the Eleventh In-*

ternational Electronics Manufacturing Technology Symposium, pp. 440-445, 1991.

[5.19] M. M. Salatino and R. C. Bracken, "Assembly Choices in Multi--Chip Module Fabrication, The Harris Digital Drop Receiver DDR-1," *Proceedings of the 1st International Conference on Multichip Modules*, pp. 74-82, 1992.

[5.20] J. Shiao and D. Nguyen, "Performance Modeling of a Cache System with Three Technologies: Cyanate Ester PCB, Chip-on-Board, and Cu/PI MCM," *Proceedings of the IEEE Multichip Module Conference*, pp. 134-137, 1992.

[5.21] P. A. Sandborn, "Technology Application Tradeoff Studies in Multichip Systems," *Proceedings of the 1st International Conference on Multichip Modules*, pp. 150-158, 1992.

[5.22] P. A. Sandborn, H. Hashemi and L. Bal, "Design of MCMs for Insertion into Standard Surface Mount Packages," *Proceedings of the National Electronic Packaging and Production Conference (NEPCON-West)*, pp. 651-660, 1993.

[5.23] S. Rao, B. Haskell, and I. Yee, "Trade-Off Analysis on Cost and Manufacturing Technology of an Electronic Product: Case Study," Proceedings of the Second International Workshop on The Economics of Design, Test, and Manufacturing for Electronic Circuits and Systems, 1993.

[5.24] H. B. Bakoglu, *Circuits, Interconnections, and Packaging for VLSI*, Addison-Wesley Publishing Company, 1990.

[5.25] D. W. Dobberpuhl, et al, "A 200-MHz 64-b Dual-Issue CMOS Microprocessor," *IEEE J. of Solid-State Circuits*, vol. 27, no. 11, pp. 1555-1567, November, 1992.

[5.26] P. H. Dehkordi and D. W. Bouldin, "Design for Packagability: The Impact of Bonding Technology on the Size and Layout of VLSI Dies," *Proceedings of the IEEE Multichip Module Conference*, pp.

153-159, 1993.

[5.27] M. Abadir, A. Parikh, L. Bal, P. Sandborn, and C. Murphy, "High Level Test Economics Advisor (Hi-TEA)," *Proceedings of the Economics of Design, Test, and Manufacturing Workshop*, 1993.

List of Symbols

a	fin length
$area_{core}$	core die area
A	fraction of nets which are nearest neighbor routed (Section 3.3.1)
A	area (Section 3.3.3 and 3.3.6)
A_a	actual availability
A_d	cross-sectional area of a duct
A_f	fin surface area
A_i	inherent availability
b	feature size parameter (length) (Section 3.3.1)
b	fin height (Section 3.3.3)
b	unusable wafer border width (Section 3.3.6)
B_p	bond pad pitch
c	wiring correlation constant (dimensionless)
c_p	specific heat
$count_{input}$	number of identical devices at the input to a process step
C	capacitance
C_L	load capacitance on a transmission line
C_m	mutual capacitance per unit length of a coupled transmission line
C_o	capacitance per unit length of a transmission line
C_{ox}	capacitance per unit length of a gate oxide in an MOS transistor
d	diameter (Section 3.3.3)
d	duty cycle (Section 3.3.4)
d_w	wafer diameter
D	average defect density on a wafer

D_h	hydraulic diameter
E_a	activation energy
f	average fanout of a chip's I/O (Section 3.3.1)
f	friction factor (Section 3.3.3)
f_m	fraction of metal in a reference plane
F_i	the fraction of the interconnect wiring that can be used for routing
F_p	chip footprint dimension (length)
F_v	via use efficiency
g	fin separation
g	acceleration of gravity (footnote 3 in Section 3.3.3)
$grid_x$, $grid_y$	number of grid cells in the x and y direction (for dc drop analysis)
h	heat transfer coefficient
I_c	interconnect capacity or connectivity
I_{ground}	current flowing in a ground lead
I_{max}	peak switching current
I_{signal}	current flowing in a signal lead
k_t, k	thermal conductivity (Section 3.3.3)
k	gate depth of a circuit (Section 3.3.4)
k	Boltzmann constant (Section 3.3.5)
K_p	proportionality constant in Rent's rule (dimensionless)
l	channel length in an MOS transistor
$length_{pad}$	length of a bond pad on a die
$\mathcal{L}$	transmission line length
L	length of a board or module
L	inductance
L_{avg}	average interconnect net length in a module
L_{bond}	bond length
L_{chip}	length of a bare die
L_{eff}	effective inductance of a ground lead
L_m, L_{mutual}	mutual inductance per unit length of a coupled transmission line
L_o, L_{self}	(self) inductance per unit length of a transmission line
m	mass flow rate
M	maximum number of repairs before scrapping

n	average number of pins per net (Section 3.3.1)
n	number of identical ducts in a heat sink or number of thermal vias under a die (Section 3.3.3)
n	harmonic number (Section 3.3.4)
n	number of simultaneously switching drivers (Section 3.3.4)
n	number of components (Section 3.3.5)
$n_{remaining}$	number of I/O remaining to be escape routed
$n_{thermal\ vias}$	number of thermal vias under a die
n_{top}	the number of I/O that can be escape routed on the top layer of an interconnect
N	number of crossings
N_c	total number of I/O per chip
N_{chip}	number of chips in a module
N_{cir}	number circuits per chip
N_{cross}	number of nets that cross a section drawn through a board or module
N_g	number of gates in a system
N_L	number of packaging levels
N_s	number of signal and control I/O in a system
N_{sc}	number of signal and control I/O per chip
N_{se}	number of signal and control I/O leaving a module
Nu	Nusselt number
N_w	number of signal wiring layers in an interconnect
p	degree of parallelism of a logic complex (dimensionless) (Section 3.3.1)
p	pitch (Section 3.3.3)
p_c	power dissipation per circuit
pm	frequency of preventative maintenance actions
p_{top}	number of lines (tracks) allowed between pads on the top layer of an interconnect
$pitch_{pad}$	center-to-center pitch of bond pads on a die
P	pressure (Section 3.3.3)
P	entity center-to-center pitch (Section 3.3.4)
P_c	power dissipation of a chip
P_{cs}	probability that a net crosses a section drawn through a board or module

P_{c_i}	power dissipation of the ith cell (dc drop analysis)
P_m	power dissipation of a module
P_o	reference power
P_1, P_2	fraction of signal pins on side 1 or 2 of a cross-section drawn through a board or module
q	heat flux
q_A	maximum heat that can be removed from a chip per unit area
Q	heat flow
r	radius
rh	relative humidity
R	resistance (Section 3.3.4)
$R_c(t)$	probability of correct component operation at time t
Re	Reynolds number
R_{ext}	external thermal resistance
R_{int}	internal thermal resistance
R_m	average interconnect length (in units of chip footprint size, F_p)
R_o	dc resistance per unit length of a transmission line
R_{on}	on-resistance of an output stage
$R_s(t)$	probability of correct system operation at time t
R_t, R	thermal resistance (degrees C/W) (Section 3.3.3)
s	minimum spacing between bonds or leads on adjacent components, or minimum spacing between die on a wafer
S:G	signal to ground I/O ratio
t	material thickness or fin thickness (Section 3.3.3), plane thickness (dc drop analysis)
t	time (Section 3.3.4)
$t_{charging}$	RC charging delay
t_{ct}	thickness of a connector trace
t_{cycle}	cycle time
t_m	mission time
t_{on}, t_{off}	signal on and off times
t_r, t_f	signal rise and fall times
t_{tof}	time of flight delay

T	temperature
T_c	number of tracks per channel in an interconnect
$T_{coolant}$	coolant temperature
$T_{coolant\ in}$	inlet coolant temperature
T_j	junction temperature
T_p	period
U	power-delay product
v	wiring correlation constant dependent on the type of module connector
V	cooling fluid velocity (Section 3.3.3)
V	voltage level (Section 3.3.4)
V_c	voltage across a capacitor in an RC circuit
V_{dd}	gate bias
V_{noise}	switching noise magnitude
V_p	via or hole pitch
V_{supply}	supply voltage for a chip
V_t	threshold voltage
w	channel width in an MOS transistor
w_{ct}	width of a connector trace
w_x, w_y	width of a module in the x and y directions
$width_{pad}$	width of a bond pad on a die
W_{chip}	width of a bare die
W_p	wiring pitch
$yield_{assembly}$	yield of an assembly step
$yield_{input}$	yield of a component at the input to an assembly step
$yield_{module}$	module yield
$yield_{output}$	yield of a component at the output of an assembly step
$yield_{repair}$	yield of a repair step
$yield_{wd}$	yield of die on a wafer
Z_o	characteristic impedance of a transmission line
Z_{oe}	even mode impedance of a coupled transmission line system
Z_{oo}	odd mode impedance of a coupled transmission line system
α_1, α_2	coefficients which depend on the signal level to which delay is measured

β	Rent's constant (dimensionless)
ε_{eff}	effective relative dielectric constant
ε_o	permittivity of free space
η	fin efficiency
κ	fractional increase in the core die area necessary to accommodate redistribution of one I/O to the periphery of the die
λ_b	component base failure rate
λ_c	component failure rate
λ_s	system failure rate
μ	surface mobility
μ_o	permeability of free space
π_i	acceleration factor
ρ	mass density (Section 3.3.3)
ρ	resistivity (Section 3.3.4)
ρ_p	allowed power dissipation density (power dissipation per unit area)
ρ_{via}	via or hole density (number of vias or holes/unit area)
τ_{wgd}	weighted gate delay
υ	viscosity
ϕ_n	phase of the nth harmonic from a Bode phase plot
ω	frequency (radians)

Index